DATE DUE

GAYLORD   PRINTED IN U.S.A.

Stardust

Stardust

Supernovae and Life—
The Cosmic Connection

John Gribbin
with Mary Gribbin

 Yale Nota Bene
Yale University Press   New Haven and London

Ay, for 'twere absurd

To think that nature in the earth bred gold

Perfect i' the instant; something went before.

There must be remote matter

– Jonson, *The Alchemist*

First published as a Yale Nota Bene book in 2001.
Copyright © 2000 by John and Mary Gribbin.

For information about this and other Yale University Press publications, please contact:

U.S. office   sales.press@yale.edu
Europe office   sales@yaleup.co.uk

Printed in the United States of America

The Library of Congress has catalogued the hardcover edition as follows:
Gribbin, John R.
Stardust : supernovae and life : the cosmic connection / John Gribbin with Mary Gribbin.
    p.   cm.
Includes bibliographical references and index.
ISBN 0-300-08419-6 (alk. paper)
1. Cosmochemistry. 2. Supernovae. I. Gribbin, Mary. II. Title.
QB450.G75 2000
523'.02—dc21   00-035944

ISBN 0-300-09097-8 (pbk.)

A catalogue record for this book is available from the British Library.

10 9 8 7 6 5 4 3 2 1

Contents

Preface

You Read It Here First!

In January 2001, scientists from NASA's Ames Research Center and the University of California, Santa Cruz, surprised many of their colleagues and created headline news by announcing the results of experiments carried out in laboratories here on Earth which produced complex organic molecules under conditions resembling those which exist in interstellar clouds of gas and dust. In these experiments, a mixture of the kind of icy material known to exist in those clouds (composed of water, methanol, ammonia and carbon monoxide frozen together) was kept in a cold vacuum and dosed with ultraviolet radiation. Chemical reactions stimulated by the radiation (typical of the kind of radiation from young stars which zaps real interstellar clouds) produced a variety of organic compounds which, when immersed in water, spontaneously created membranous structures resembling soap bubbles. All life on Earth is based on cells, bags of biological material encased in just this kind of membrane. The implication of this work is that space is filled with chemical compounds which can easily give a kick-start to life if they land in a suitable environment, such as on the surface of the Earth. The comets which proliferate in the outer part of our Solar System and occasionally pass through the inner regions near the Earth are known to be made of almost pristine interstellar material left over from the formation of the Sun and planets out of one of these interstellar clouds. It seems very likely, therefore, that any planet like the Earth will be seeded with the raw materials necessary for life almost as soon as it forms.

The discovery was headline news at the time, and seems to have taken the researchers themselves by surprise. In the issue of *The Independent* dated 30 January 2001, Lou Allamondola, the leader of the team, was quoted as saying "we expected ultraviolet radiation would make a few molecules that might have some biological interest, but nothing major. Instead, we found that this process transforms some of the simple chemicals that are very common in space into larger molecules which behave in far more complex ways, which many people think are critical to the origin of life." But these researchers, and the journalists so eager to write their headlines, might have been less surprised by the discovery, and the implications for the origin of life on Earth, if they had kept up-to-date with the story told in this book. There is nothing new, as you will see, with the idea that complex organic material is built up in space from simple atoms and molecules by the action of ultraviolet radiation; nor is there anything new in the suggestion that these precursors of life may have been brought down to Earth by comets. Indeed, the story is better presented as an example of the classic scientific method at work. The work described in the original edition of this book essentially provided the basis of a prediction, that complex molecules (and even cell-like structures) must exist in the kind of interstellar clouds from which planetary systems like our own form. That prediction has now been borne out by experiment, which elevates the status of the ideas described here from that of a mere hypothesis to that of a fully fledged theory. There now seems very little room to doubt that life must be a common occurrence across the Universe (which is not quite the same thing as saying that intelligent life is common). To find out why, read on.

Introduction

Our Place in the Universe

Life begins with the process of star formation. We are made of star-dust. Every atom of every element in your body, except for hydrogen, has been manufactured inside stars, scattered across the Universe in great stellar explosions, and recycled to become part of you. The hydrogen is primordial material, produced in the Big Bang, along with helium (there is no helium in your body). Hydrogen and helium together formed the raw material for the first generation of stars, some 12 billion years ago, but everything else has been built up by nuclear fusion in stellar furnaces.

The drama and appeal of this discovery was brought home to me by the response I get whenever I give popular talks about astronomy and mention this well-established link between ourselves and the stars. I have often been asked, "Why don't you write a book about that?" My response has been, "I will—when the time is ripe." Now, the time is ripe. I decided to write this book in the wake of a wave of discoveries of planets orbiting around other stars in our Milky Way Galaxy. If other planets exist — perhaps other Solar Systems like our own — the possibility of finding other forms of life in the Universe is greatly increased. But it seems to me that before we get too deeply involved in speculations about extraterrestrial life, we ought to understand our own place in the cosmos. I hope to convince you that we are a natural product of the Universe we live in, and that therefore it is natural to expect other life forms, perhaps rather like us, to exist elsewhere in the cosmos.

Since the Sun, the Earth, and everything on Earth (including ourselves) are now seen as the natural by-product of the existence of stars and the Milky Way Galaxy in which we live, the implication is that it is highly likely that other planets like the Earth, and other life forms like us, exist. But I do not intend to speculate on exactly how life got started, or precisely where we might find it outside our own planet. The strength of the story I have to tell is that it deals almost entirely with facts, not speculation.

The story really begins in the 1920s, when astronomers began to appreciate that a star like the Sun is indeed, even today, largely made of hydrogen and helium — before that, they had assumed that stars were made of much the same sort of material as planets like the Earth, which is rich in iron, the most stable element.

Beginning in the 1920s, the story of how we are made of stardust, and are therefore the children of the stars, involves the understanding of how stars themselves work that was developed over the next few decades. It is no coincidence that this understanding developed when it did, because it involved both the special theory of relativity and quantum physics, ideas which were themselves new to science in the early twentieth century. In the nineteenth century, the fact that stars stayed hot at all was one of the greatest puzzles confronting not just astronomy, but physics.

The story of stardust is also inextricably linked with the idea of the birth of the Universe in the Big Bang. In the 1940s George Gamow showed that the Big Bang could have produced hydrogen and helium, and although his detailed explanation of how the heavier elements were built up from these primordial elements later proved wrong, he claimed he was still happy because the hydrogen and helium together accounted for 99 percent of the stuff of the Universe known to observational astronomers at that time. In the 1950s, a team led by the British astrophysicist Fred Hoyle showed how the other 1 percent of everything could be manufactured inside stars, and in the 1960s Hoyle and his colleagues went back to the Big Bang idea and explained subtle details of the processes in which the raw material of the first generation of stars was manufactured. In the 1970s and 1980s, astronomers focused on details of the behavior of supernovae, the stellar explosions which distribute the raw materials of new generations of stars, planets, and people across the cosmos — and now, they are simulating certain aspects of these events in particle accelerators here on Earth.

This is the story I have to tell. It focuses on the key issue in the relationship between ourselves and the Universe, the question of how the chemical elements we are made of were manufactured inside stars and distributed through space. Inevitably, because of the intimate relationship between the lives of stars and the life of the Universe, there is some overlap with my earlier books about cosmology, especially *The Birth of Time*. I hope this will not be too obtrusive for anyone who has read those books; to me, the way the pieces of the astronomical jigsaw puzzle fit together to make a seamless whole is itself a major discovery, which shows that the whole enterprise of science is very much on the right track in unraveling the mysteries of the Universe.

At the heart of the present story, you will find the supernovae, great stellar explosions in which a single star briefly shines as brightly as a hundred billion ordinary stars like the Sun. And at the end of the story, a little speculation about the relationship between life and the Universe — or universes.

To set the scene, though, we need to know just a little background to the story of our place in the Universe. If you already know about the structure of the Solar System, what planets are and how astronomers think they formed, you will miss nothing by skipping the rest of this introduction and moving on to the meat of the book. But for those new to the story, or anyone who wants a reminder, read on.

The Sun is a star, one of a couple of hundred billion similar stars which together form a disk-shaped system called the Milky Way Galaxy. In round terms, the disk of the Milky Way is about a hundred thousand light years across, which literally means that light takes one hundred thousand years to cross it (even though light travels at three hundred thousand kilometers per second). The Sun and its family of planets — the Solar System — orbit round the center of the Milky Way at a distance roughly two-thirds of the way out toward the edge of the disk, taking a couple of hundred million years to complete each orbit. It seems to be an ordinary star in an ordinary part of the Milky Way — and, although it is not central to my story, the Milky Way Galaxy seems to be an ordinary galaxy, one of several hundred billion similar objects scattered throughout the entire visible Universe.

To put the size of the Sun in perspective, its diameter is a little more than one hundred times that of the Earth, so its volume (proportional to the cube of its diameter) is a bit more than a million times that of the Earth. Like all stars, the Sun shines because nuclear reactions are going on in its interior, generating energy (more — much more — on this later).

The Sun is accompanied by a family of planets and smaller objects, all orbiting round the Sun (held in orbit by the Sun's gravity) and together making up the Solar System. There are four relatively small rocky planets, which orbit relatively close to the Sun — Mercury, Venus, Earth, and Mars. Then there is a region where millions of chunks of cosmic rubble orbit in a band or ring around the Sun. This is known as the asteroid belt; there are at least a million asteroids each bigger than one kilometer in diameter, and uncounted numbers of smaller pieces of debris. Beyond the asteroid belt, there are four large gaseous planets — Jupiter, Saturn, Uranus, and Neptune. A ninth object, Pluto, is usually graced with the name "planet," even though it is only a ball of ice just two-thirds of the size of our Moon. Beyond even the orbit of Pluto, the Sun is surrounded by a swarm of icy objects called comets.

Don't worry about the details. There are just two important features of this system to take on board for now. First, there are small rocky planets close to the Sun and large gaseous planets farther out from the Sun. Second (and much more important), everything out to Pluto (and actually a bit farther) orbits round the Sun in the same direction, and in the same plane, like runners running in their own lanes round a cosmic running track. Most moons orbit their planets in this same sense, and most planets rotate on their axes in the same direction. This is extremely powerful evidence in support of the idea that the whole Solar System formed from a cloud of gas and dust in space that collapsed under its own weight. As it shrank, it spun faster and faster, like a spinning ice skater pulling in her arms, and leftover material settled into a disk moving around the young star in the same direction that it was rotating. Eventually, planets formed in this dusty disk; but I really do mean that they formed from "leftover" material. Some 99.86 percent of the mass of the Solar System today is concentrated in the Sun itself, and two-thirds of the rest is locked up in the giant planet Jupiter. Everything else put together (including the Earth) makes up less than 0.05 percent of the mass of the whole Solar System.

If this picture of how planets and solar systems form is correct, there ought to be disks of dusty material around many of the young stars in the Milky Way itself. Such disks are hard to detect, because they do not shine like a star, but only glow dully as they are warmed by the heat of the star in the middle or scatter and reflect some of its light away into space. Although there is indirect evidence of dusty disks around several stars (for example, from the way some of the starlight is obscured), until a few years ago just one of these disks, around a young star called Beta Pictoris, had been imaged using ordinary visible light (the disk was first photographed in 1984). But in 1998 astronomers were able to obtain images of such disks around three other young stars, using infrared detectors.

The infrared is the part of the electromagnetic spectrum with wavelengths slightly longer than red light. Our eyes cannot see infrared radiation, but we feel it on our skin as the radiant warmth from a fire or, indeed, from a central-heating radiator. Part of the infrared band can be "seen" using suitable cameras (like the ones used in night vision systems, sensitive to infrared heat radiation) attached to optical telescopes, and part is "visible" to a special kind of radio telescope. Both techniques were used to obtain images (or maps) of the disks around two of the brightest stars in the sky, Vega and Fomalhaut, and around a less prepossessing object which goes by the name HR4796A.

Apart from the fact of the existence of the disks, the importance of these discoveries lies in the ages of the stars concerned. HR4796A is about 10 million years old, Beta Pictoris about 35 million, Fomalhaut about 100 million, and Vega about 350 million (astronomers estimate the ages of stars by comparing their computer simulations of how stars work with the observed appearances of the stars). Our Sun is about 4.5 billion years old, and evidence from the radioactive dating of rocks on Earth and samples from the Moon and meteorites (bits of cosmic debris that have fallen to Earth) suggests that planet formation was completed within about a hundred million years of the Sun's "switching on." So these systems are exactly the right age to be systems like our Solar System in the process of formation.

Incidentally, in 1998 Vega became widely known outside the ranks of professional astronomers thanks to the movie *Contact,* in which it was the source of signals from an intelligent alien race; but Vega is actually an extremely unlikely place to find intelligent life, both because the system is too young for intelligence to have evolved there and because all the activity going on in the dusty disk, with bits of debris banging into one another and crashing onto young planets in the process of formation, would make it a very unhealthy place for any space-faring species to visit.

All these disks are much bigger than our Solar System. Planetary astronomers measure distances in terms of the astronomical unit, or AU, which is the average distance from the Earth to the Sun (almost exactly 150 million kilometers). Neptune orbits the Sun at a distance of 30 AU (thirty times farther from the Sun than we are), so in round terms the planetary part of our Solar System is about 60 AU across, equivalent to the diameter of Neptune's orbit. The dusty disks seen around young stars are typically a few hundred astronomical units across — but, significantly, in both HR4796A and Fomalhaut there is a clear inner region. The cleared space around the central star in

Fomalhaut is 60 AU across, and the hole in the disk around HR4796A is about 70 AU across. Beta Pictoris itself, the archetypal disk star, has a smaller hole, about 20 AU across. And in all the disks seen so far, there is nowhere near as much matter as exists in all the planets of the Solar System put together—the dust is spread very thin. All this suggests that planet formation may have cleared the inner regions of the disks, as particles of dust have stuck together and accumulated to make larger objects. Some of the material in the outer parts of the extended disks may go to make comets, but a great deal is in the process of being blown away into interstellar space by the heat of the young star at the center of the system.

As well as the dusty disks around some stars (and more continue to be discovered, so any list I would give here is bound to be incomplete by the time you read this), in the late 1990s astronomers at last began to turn up evidence of planets around other stars. Most of this evidence comes from painstaking studies of the way relatively nearby stars move across the sky. In most cases, these movements are measured by changes in the spectrum of light from the star, caused by its shifting backward and forward as it is tugged by the gravity of a planet in orbit around it. We cannot see the "wobble" in the star's motion directly.

There was a claim that one of these stars had been seen to wobble from side to side, by a tiny amount. These claims now seem to have been premature, but to give you some idea of how difficult it is to measure such changes, in a typical case the rhythmic displacement of the star in the sky would be equivalent to measuring a sideways shift across a distance equivalent to the width of a single human hair viewed at a distance of one and a half kilometers. It is hardly surprising that the claims have been disputed! But there is no dispute about the stellar wobbles revealed by spectroscopy. These wobbles are interpreted as a result of the gravitational tugging exerted on the stars by giant planets, orbiting around them and rhythmically pulling them from side to side (in most of the cases studied so far, you need planets even bigger than Jupiter to explain the wobbles). There is one case in which a tiny dimming in the light from a star might be explained by a planet only a little bit bigger than the Earth moving across in front of it, in a so-called transit, like a miniature eclipse. The star is called CM Draconis, a name worth remembering in the hope that more observations confirm these results. In round terms, the number of "extrasolar" planets claimed by astronomers reached a dozen in the second half of 1998, and although the experts are still arguing about some of these interpretations, this is enough that it seems highly unlikely that they will all be explained away in some other fashion.

Writing at the beginning of 1999, it seems safe to say that we have definite proof that other planets exist, and we also have direct observations of the dusty disks in which we expect planets to form. There is one other compelling piece of evidence in all this work. From spectroscopy, which analyzes the light emitted from stars and identifies the characteristic fingerprints of different elements from the pattern of bright and dark lines they make in the spectrum (like a kind of cosmic barcode), we know what stars are made of. The oldest stars, which formed from the primordial material produced in the birth of the Universe in a Big Bang, contain only hydrogen and helium and very tiny traces of a few other light elements. Younger stars, made more recently from material which has been partly processed inside other stars and recycled to make new stars (as I discuss later), contain a lot more in the way of heavier elements. With a cavalier disregard for the subtleties of chemistry, astronomers usually lump all the elements except hydrogen and helium together under the label "metals" (so, to an astronomer, oxygen is a metal). And all the stars which have been found to have planets orbiting around them are relatively rich in metals — they are all stars which have been made from thoroughly recycled material. This is as you would expect, since how can a planet, made of things like carbon and sulfur and silicon and oxygen and nitrogen, form from a cloud of gas which contains only hydrogen and helium — which is how the first stars must have formed? The dusty disks in which planets form are literally stardust, the product of the activity of previous generations of stars. And this is where my story really begins. It is the story of the origin of the 0.05 percent or less of star stuff that goes to make planets and people. There would be no planets like the Earth, and no life forms like us, if there were no clouds of gas in space laced with that tiny trace of dusty debris produced by previous generations of stars.

I'm not going to try to explain the details of exactly how a dusty disk like the ones around Fomalhaut and Beta Pictoris is associated with the formation of a planet like the Earth, because astronomers do not yet know exactly how this happens; and I'm certainly not going to try to explain in detail how life originated (though I cannot resist a modest speculation). But I am going to tell you, briefly, how hydrogen and helium were manufactured out of pure energy in the Big Bang itself. And I am going to tell you in much more detail where the stardust that those disks are made of came from. There is no helium in your body, and no need to look inside stars for the origin of the hydrogen, but I'm going to tell you about the stellar origin of every atom of every other element in your body. The story of what "life as we know it" is all about is in large measure the story of the Universe we live in, because life and the Universe are inextricably intertwined.

Life and the Universe

This book explains the relationship between life and the Universe, from the Big Bang to the arrival of the molecules of life on the surface of the Earth. It is a complete and self-consistent story, describing our cosmic origins from stardust. But it is not necessarily the whole story of life and the Universe, and before delving into the details, I want to describe briefly some of the more intriguing current ideas that may, if they are proved correct, take us beyond the story so far. The caveat is that "intriguing" doesn't necessarily mean "correct." But science progresses by making reasonable speculations, then testing those speculations to see how well they stand up. And in a book which claims to offer the best available scientific evidence for our own origins, it would be derelict not to make it clear just how science arrives at these profound conclusions.

One particular speculation about the relationship between life and the Universe has the merit of being relevant to my story in addition to highlighting the scientific method at work. It is actually a rather old idea that has recently been revived and improved in the light of present astronomical knowledge. It's especially interesting because it shows how scientific ideas can go in and out of fashion—and then back in again—as new discoveries are made and opinions change. In fact, as is so often the case in science, the first person to put the idea forward was way ahead of his time. In 1871, William Thomson (who later became Lord Kelvin) pondered the mystery of the origin of life on Earth in his presidential address to the British Association for the Advancement of Science. He made an analogy with the way life appears on a newly formed volcanic island, saying that "we do not hesitate to assume that seed has been wafted to it through the air, or floated to it on raft—we must regard it as probable in the highest degree that there are countless seed-bearing meteoric stones moving about through space. If at the present instant no life existed upon this Earth, one such stone falling upon it might, by what we blindly call natural causes, lead to its becoming covered with natural vegetation."

Thomson's comments are especially interesting as an echo of their times—they came just a dozen years after Charles Darwin and Alfred Wallace published the theory of evolution by natural selection, a key point of which was the way life forms appeared on isolated islands and evolved there into new species. The meteoric reference also echoes Thomson's own interest in how the Sun stayed hot—in a step toward his idea that the Sun could release heat by contracting slowly under its own weight, Thomson had investigated theoretically the possibility that it might be kept hot by a continual rain of meteoric debris falling onto its surface. But Thomson seldom gets any credit for his ideas about the origin of life on Earth, and although he deserves a mention, it is probably fair that his theories should be relegated to a footnote in scientific history, since he never developed them fully (or at all!) and left them as pure speculation.

The story really begins in 1907, with a suggestion made by the Swedish chemist Svante Arrhenius. Arrhenius was a good enough chemist to have won the Nobel Prize in 1903 for his work on electrolysis, and the breadth of his scientific thought is reflected in the fact that he was one of the first people, in 1905, to express concern about the prospect of global warming caused by a buildup of carbon dioxide in the atmosphere (the greenhouse effect) as a result of burning fossil fuels. His interest in the workings of the atmosphere of our planet led directly to his speculations about the origin of life on Earth, after he realized that it would be possible for microorganisms (things like bacteria) to be carried high into the atmosphere, where they might escape into space and be pushed outward from the Solar System by the pressure of the Sun's radiation. It is known that some forms of such microorganisms can remain dormant for long periods of time in hostile environments (particularly under arid conditions) and then be brought back to active life when their essential requirements (particularly water) are once again available. Perhaps, he reasoned, they could even cross the desert of interstellar space in this dormant state, reviving when they fell upon another Earth-like planet.

But why should this be a one-way process? If living spores from Earth could escape into space in this way, Arrhenius pointed out, then spores from other planets, orbiting round other stars, could also escape into space; life on Earth might therefore have descended from such interstellar travelers which entered the atmosphere of the Earth when our planet was young. This hypothesis for the origin of life on Earth was called "panspermia," meaning "life everywhere," and it fitted in rather neatly with the image of the Universe people had at the beginning of the twentieth century. Arrhenius knew nothing of Thomson's speculation, and in any case he offered a properly worked-out hypothesis, trying to explain not only how life could have got from rocks in space onto a planet but how it could have got off a planet and into space. He deserves the pride of place that he is usually given in the story of panspermia.

At that time, what we now know as the Milky Way Galaxy was thought to be the entire Universe. Astronomers already knew that individual stars were born, lived, and died within the Milky Way, but it was thought that the "Universe" itself was essentially eternal and unchanging — as an analogy you might think of an ancient wood that has existed since time immemorial, even though every tree in the wood has been replaced by new trees very many times. The key feature of this picture of the Universe is that there was no origin, so the question of how the Universe began did not have to be addressed. On the other hand, there was clearly a question about how life on Earth had originated, since radioactive dating techniques had begun to put a date on the age of the Earth by the time Arrhenius was pondering this puzzle. But by moving the origin of life off the Earth and out into what was then thought to be an eternal Universe, Arrhenius "solved" the puzzle by removing it from consideration altogether. If the Universe was eternal and essentially unchanging, even though generations of stars ran through their life cycles within the Universe, then it seemed reasonable to argue that life had always existed in the Universe and had spread from old planets to new ones as part of the cycle of the generations. And in an infinitely old Universe, even if life did have to emerge by chance processes, there would be infinite time available to do the job, and then infinitely more time for life to spread from its planet of origin to populate the entire Universe. This was entirely sensible reasoning, given what was known about the Universe in the first decade of the twentieth century.

Even so, it was not taken very seriously, and as our understanding of the stars, the Milky Way, and the Universe at large developed over the next half century (a story elaborated on in the heart of this book), most of the people who thought about the problem of the origin of life worried about how to make complex organic molecules from simple chemicals such as methane and ammonia under the conditions that were thought to have existed on the early Earth — as we shall see, it was not until the late 1960s that radio astronomy began to reveal the richness of interstellar chemistry.

It was also in the 1960s that interest in the panspermia idea stirred once again — though this actually happened before the discovery of complex organic molecules in space. Part of the impetus for this interest came from balloon flights that carried unmanned instrument packages to great heights in the stratosphere and showed that microorganisms do indeed float around in the upper atmosphere. But the key calculations were carried out by the American astronomer Carl Sagan, when he collaborated with the Russian Iosef Shklovskii on the epic book *Intelligent Life in the Universe,* first published in 1966 (but still well worth reading). Instead of simply speculating about the fate of such microorganisms, Sagan actually calculated the effect of solar radiation on particles of different sizes (something Arrhenius could not do, of course, because in the early 1900s there was insufficient information about the Sun and the interplanetary environment).

Because gravity pulls particles in toward the Sun, and the radiation pressure pushing them outward is rather feeble, it turns out that only very small particles can be blown away from the orbit of the Earth — microbes less than about half a millionth of a meter in diameter. This is intriguing on two counts: first, because there are living microorganisms just that size; and second, because this is also just the size of the particles of dust that have now been detected in interstellar clouds. Such a bacterial particle departing from the Earth would pass the orbit of Mars in a few weeks, Jupiter in a few months, escape from the Solar System in a few years, and perhaps mingle with an interstellar cloud within a million years. Although the original panspermia idea envisaged the microorganisms drifting down into the atmospheres of newly formed planets, a more modern interpretation would see them as becoming part of the material out of which new planetary systems form. But there is a snag, which Sagan was quick to point out.

As soon as the microorganisms leave the Earth's atmosphere, they are exposed to ultraviolet (UV) radiation from the Sun and also to particles, such as protons and electrons, that form part of the solar wind (solar cosmic rays). Even the most resistant bacteria around on Earth today would be killed by the UV radiation within a day of leaving the Earth, and even if there existed an organism that was impervious to solar UV, it would be killed by the cosmic rays before it got out of the Solar System.

There's another problem, at least in the original version of this idea. If microorganisms about 0.5 millionths of a meter in size are blown outward from the orbit of the Earth, certainly nothing in this size range could have fallen and landed on the young Earth, even if it had escaped from a similar planet somewhere across the Universe. This led Sagan and Shklovskii to discuss the possibility of the seeds of life arriving on planets farther out from their parent star — planets like Jupiter or Saturn. This rather begs the question of how life originated on Earth; but in any case this problem doesn't really arise if we envisage the microorganisms as becoming integrated into interstellar clouds from which new planetary systems form, because they would be carried down onto young planets by cometary impacts, part of the natural process of planet formation that I describe in Chapter 9.

By the early 1970s, Sagan was convinced that the panspermia idea would not work, because the environment of space is too hazardous for the kind of living things that could escape from the Earth today. But at about the same time, Francis Crick,[1] an eminent British biologist, was becoming convinced that the astronomical and geological evidence allowed too little time for life to have evolved from scratch on the Earth itself—there is clear geologic evidence that life existed on Earth less than 600 million years after the planet formed, and biologists such as Crick see no way it could have emerged from a mixture of simple chemicals in that time. While the astronomer was rejecting panspermia on biological grounds, the biologist was about to begin promoting it on astronomical grounds. Together with the American Leslie Orgel, Crick developed a variation on the theme, which he called "directed panspermia," arguing that the Earth had been deliberately seeded with life, in the form of microorganisms (essentially bacteria) carried across space inside an alien spacecraft safely shielded from cosmic radiation.

This would not necessarily have to be interpreted as a deliberate attempt to seed the Earth—we very nearly have the technology today to send small unmanned probes out in more or less random directions, carrying bacteria to be dumped on any planet they encounter. It may seem an inelegant way for life on Earth to have got started, but it was one step better, perhaps, than the proposal put forward by the astronomer Tommy Gold, who suggested, only slightly tongue in cheek, that all of life on Earth might have descended from the organic

1. Who shared the 1962 Nobel Prize with James Watson for the discovery of the structure of DNA.

debris left behind on the planet by some aliens who stopped by for a picnic!

Since the 1970s, though, the pendulum has swung again, and a new variation on the panspermia theme has suggested how, in spite of the radiation problem, microorganisms could, after all, escape by natural means from a planet like the Earth and cross interstellar space to infect other planets with life. Jeff Secker, at Washington State University, working with Paul Wesson and James Lepock, at the University of Waterloo, in Canada, took another look at the problem of the survivability of interstellar spores and this time took account of the way a star like the Sun changes as it ages and becomes what is known as a red giant. Their first step was to imagine the living (or dormant) microorganisms being protected by being embedded in grains of dust. This only partially solves the radiation problem today and also makes the particles heavier, so it is harder for them to be blown out of the Solar System. When the Sun becomes a red giant, the intensity of ultraviolet radiation from its surface will be much less, but its overall brightness will increase (thereby increasing the pressure of radiation pushing tiny grains outward from Earth orbit) and the strength of the solar wind of material blowing away into space will be much greater — as we shall see, red giants eject a lot of material into space, and life-bearing material from a planet in orbit around the red giant could well be included in this. Once again, it is easy to see how the living material could end up in interstellar clouds from which new planetary systems form. And it doesn't even have to be *living* (or dormant) material. In 1996 Secker and his colleagues pointed out something that everyone else seems to have missed: even what they call "inactivated" biological material (in the form of fragments of molecules such as DNA, the remains of once-living material broken apart by cosmic rays) could, if introduced to a suitable planet, "enhance the chance that life will evolve there, and could possibly explain the (apparently) rapid evolution of early life on Earth." Once again, though, like everyone who has promoted panspermia in its various guises, Secker and his colleagues are thinking in terms of the biological material falling onto a preexisting planet and have missed the point that it is much more simple for the material to mingle with the stuff from which planets form in the first place.

From their point of view, the good news is that since a star like the Sun becomes a red giant only after 10 billion years or so of its existence as a stable star, much the same state that we see it in today, there is ample time for life to evolve on a planet orbiting around such a star. The first time this happened, it could indeed have taken much more than 600 million years for life to emerge. Then, panspermia does the rest. The bad news is that if such a long time span is needed for life to emerge once, it must have happened on a planet formed a very long time before the formation of the Solar System. The Solar System formed about 4.5 billion years ago, and this argument would seem to suggest that the whole process of the emergence of life, followed by panspermia occurring after the parent star had become a red giant, would have taken much longer. Even if we imagine that the star involved was a little more massive than the Sun (which would mean that it ran through its life cycle a little more quickly than the Sun), we still have to allow billions of years for it to become a red giant, because the whole point of the argument is that you need those billions of years for the first living things to evolve. It is no good invoking, say, a star with three times the mass of the Sun, which runs through its life cycle in just 500 million years, because if life can evolve on a planet around such a star in that short a span of time, it could have done so on Earth in the first few hundred million years of our own planet's existence.

So the "red giant panspermia" argument pushes back the formation of the system in which life first emerged beyond 10 billion years, uncomfortably close to the best estimates of the age of the Universe and allowing very little time for the first planetary systems to form after the Big Bang. That far back in time, the stars had had little opportunity to synthesize the heavier elements, and it is a matter of conjecture whether there would have been enough of the right stuff around on the earliest planets to provide the raw materials of life as we know it. And it *has* to be life as we know it, because the whole argument is that we are descended directly from those first living things.

For my money, the bottom line is that panspermia could work, but that, for reasons spelled out in this book, neither "natural" nor "directed" panspermia is needed to explain the presence of life on Earth. Both ideas seem more contrived than the suggestion that the young Earth was seeded with complex organic molecules that arose through natural chemical processes going on in the interstellar cloud from which the Solar System formed — an idea promoted by Sagan in the late 1970s, in work carried out with Christopher Chyba. And if I were going to speculate further, I would wager that the complex chemistry of interstellar clouds could have gone all the way to producing genuine living molecules, rather than introduce another cumbersome step into the calculation by venturing a guess that those molecules evolved on other planets and then got ejected into space to mingle with those interstellar clouds. In the late 1990s, laboratory experiments in which the kinds of molecules found in interstellar clouds today were dosed with ultraviolet light led to the formation of a slew of organic molecules which themselves react further to produce amino acids and other biochemical molecules. Dump that lot on the Earth when it was about 600 million years old, and the difficulties which so worried Francis Crick in the 1970s disappear. As the astronomer David Buhl has put it, "the predominance of organic species [in interstellar clouds] and their similarity to the products obtained in the [laboratory] synthesis of amino-acids in the study of the origin of life suggests a very close parallel between interstellar clouds and pre-biotic chemistry."[2]

Even though I do not think panspermia is a likely explanation of our own origins, there is no doubt that in the near future we will have the capability to seed other planets with life, which raises intriguing ethical questions but goes beyond the scope of the present book. The message I hope you get from my brief account of the history of panspermia is how much progress has been achieved in the past century. In a way, the original panspermia idea was a counsel of despair. Because nobody knew how life could have originated, the speculation was that it had always existed and simply spread from place to place in the Universe. We still do not know exactly how life began — nobody had yet seen a mixture of chemicals come to life in a test tube. But, unlike Arrhenius, we do know precisely what mixture of chemicals is required for the existence of life as we know it. And we know exactly where those chemicals came from: they are the natural by-products of the processes of star formation and evolution. That is the story I am going to tell you now, beginning with the basics of what life itself is all about.

2. Quoted by Stephen Dick, *Life on Other Worlds* (Cambridge: Cambridge University Press, 1998).

Life as We Know It

What is life? Because we are alive, and we live on the surface of the Earth, it seems perfectly natural to us, without thinking about it, that life forms like us should exist on planets like the Earth. But when you do think about it, and especially when you compare superficial conditions on Earth with those of the other planets of the Solar System, at first it seems strange that the particular collection of chemicals that make up a human being should exist at all, and that there should be a planet like the Earth on which those chemicals — combinations of elements — can evolve into interesting things like peonies and people. Only when you think about it do you realize that there is something profound going on. For once, though, the unthinking first impression may be correct. It may indeed be natural for life forms like us to exist on planets like the Earth, because although we are part of something profound, the more deeply we think about the nature of life itself, the closer the links we find between ourselves and the Universe at large.

This is especially clear if you start from the bottom up and look at life in terms of the simplest chemical building blocks, the elements.

Of course, there is more to life than its chemical components. If you put all of the chemicals that make up a human being (or a peony) in a heap, you wouldn't have a living thing, just a heap of chemicals. One of the defining features of life is that it feeds off a flow of energy and uses that energy to make complex things out of simple things. In the case of life on the surface of the Earth, the flow of energy comes from the Sun; it is solar energy that turns the simple chemistry of nonlife into the complex chemistry of life. But there are also organisms that exist deep below the surface of the ocean, where they never feel the warmth of the Sun. There, the flow of energy comes from hot vents on the ocean floor that release heat from inside the Earth into the local marine environment.

This is all part of a wider pattern seen in nature, a pattern of complex (but not necessarily living) things existing where they can feed off a flow of energy. When energy flows in the right way, simple systems spontaneously arrange themselves in interesting patterns. This is called self-organization and lies at the heart of the study of complexity, one of the most exciting and intriguing areas of scientific research at the beginning of the twenty-first century. In a simple example of (nonliving) self-organization at work, if a shallow pan of oily liquid is warmed from underneath, the heat at first travels upward by conduction and the liquid does not move. Then, as the liquid gets hotter, the bottom layer starts to rise by convection, while the cooler surface liquid falls down to replace it. At first, convection is messy. But with a gentle source of heat below the pan, the convection can settle into a beautiful pattern of hexagonal cells, like a cross section through a honeycomb, the hot liquid rising along the sides of the cells and cool liquid falling down in the middle of each cell.

Although life is much more complicated than this, it also depends on a flow of energy through a system — at its most fundamental level, through a living cell. Living cells reproduce by making new cells, but the biggest question in biology is still where the first living things came from. Biologists today have a clear idea of the minimum amount of complexity required to make a living cell — some DNA, some RNA, some protein, a membrane to hold everything together, and a source of food to provide energy. Once such "minimal bacteria" existed on Earth (whether they came from outer space or emerged on the surface of the Earth itself), their evolutionary survival was enhanced by adding complexity that made the cells more efficient users of energy and better able to reproduce. But that, like the origin of the first minimal bacteria, lies outside the scope of this book. Here, I am only concerned with the origin of the material constituents of life — but I wouldn't want to give the impression that I am so much of a reductionist as to think that that is all there is to the story. Those material constituents of life are a subset of the chemical elements — and a rather restricted subset, at that. As we all learned in school, an element is the simplest substance that can take part in chemical reactions, and an element cannot be broken down into something simpler, or converted into another element, by chemical means. Just over ninety elements occur naturally on Earth; each consists of a single kind of atom — an atom is the smallest possible unit of an element. Elements (or rather atoms) can combine with one another in certain ways to form molecules, such as molecules of water. Each molecule of water, for example, contains two atoms of the element hydrogen and one atom of the element oxygen, written as H_2O. So far, so familiar. But here comes the first surprise. Although there are more than ninety elements, the visible Universe is dominated by just two kinds of atom, and the chemistry of life itself is dominated by just four.

In terms of mass, the oldest stars are made up of about 75 percent hydrogen and a bit less than 25 percent helium, with just a smattering of other elements. The visible Universe as a whole is dominated by hydrogen and helium. But we humans are made of a quite different combination of elements. This represents primordial material which has been processed inside stars and made into heavier elements. But even though this processing (and reprocessing) of stellar material has been going on for about 12 billion years, the Solar System is still dominated by hydrogen and helium. This isn't obvious to us simply because most of the hydrogen and helium is locked up in the Sun itself, while the planet we live on, the Earth, is a bit of leftover debris orbiting around the Sun.

Measuring the abundance of the elements in terms of mass (which is the same as weight for these purposes) is only part of the story, though, because atoms of different elements have different masses (sometimes even atoms of the same element have masses that differ slightly from one another, but for now I shall refer only to the most common form of each element). Each atom of helium, for example, has four times the mass of each atom of hydrogen. So, treating each atom (or atomic nucleus) as an individual particle, the Sun is made up of 90.8 percent hydrogen, 9.1 percent helium, and 0.1 percent of everything else put together — and this is very similar to the composition, determined by spectroscopy, of other stars roughly the same age as the Sun.

But we live in the planetary part of the Solar System, which formed out of the dusty disk around the young star. The lightest stuff left in the disk tended to get blown away into interstellar space by the heat of the young Sun, and hydrogen and helium are the two lightest elements of all. So the proportions of heavier elements are a little bit bigger in the planetary part of the Solar System than in the Sun itself — not because there is more of the heavy stuff but because there is less of the light stuff. In terms of mass, and taking the Solar System as a whole, hydrogen contributes 70.13 percent of the total, helium 27.87 percent, and oxygen, the third most common element by mass, 0.91 percent.

Although hydrogen and helium dominate, the fact that oxygen is the third most common element by mass in the Solar System is already a significant discovery, because oxygen plays a key role in the processes of life as we know it — so obviously important that I don't even need to explain it. If we ignore the hydrogen and helium for the moment and concentrate on the 2 percent of the Solar System that is made of other elements, the situation looks even more intriguing.

The numbers involved are so small that it is convenient to go back, once again, to counting the number of particles rather than to measuring the masses. In our part of the Universe (and, indeed, in the Universe at large), sulfur is the tenth most common element, measured in this way: for every 2 atoms of sulfur, there are 3 atoms of iron, 4 atoms each of magnesium and neon, 5 atoms of silicon, 9 atoms of nitrogen, 40 atoms of carbon, and 70 atoms of oxygen (all of which are mere traces compared with the thousands of atoms of helium and tens of thousands of atoms of hydrogen for every few atoms of sulfur or iron).[1] Apart from this top ten, there are just five other elements (aluminum, argon, calcium, nickel, and sodium) which have abundances in the range from 10 percent to 50 percent of the abundance of sulfur. Everything else is very much rarer, for reasons that I will ex-

1. Because of the way people round off these figures to whole numbers, different books may give slightly different values. Unless I say otherwise, all numbers given in this book are intended to be taken as indicative, rather than precise, values. This set is taken from William Kaufmann's *Universe* (New York: Freeman, 1988); what matters is not whether there are precisely three or four atoms of iron for every atom of sulfur, but that there is, in round terms, ten times as much carbon as neon, and twice as much oxygen as carbon.

plain later. Gold, for instance, is so rare that there are only 3 atoms of gold for every 10 million atoms of sulfur, which is one reason why gold is so valuable.

So what are we made of? What is "life as we know it" in chemical terms? You wouldn't expect there to be any helium in your body, because helium is a very unreactive gas — so much so that it is called an inert gas. It doesn't take part in chemical reactions and refuses to form chemical compounds with anything. It is also a very light gas, lighter than anything else except hydrogen — it is this combination of being inert (and therefore uninflammable) and light that makes it so desirable as the lifting gas in airships. Almost all the helium that was in our part of the dusty disk when the Earth formed has escaped into space, because it was unable to form compounds that would have locked it up in the body of the Earth. Although hydrogen is even lighter than helium, it is very happy to form compounds, and this is dramatically apparent in the oceans that cover most of the surface of the Earth, where hydrogen is combined with oxygen in the form of water.

Leaving aside unreactive helium, the two most common elements in the Solar System are hydrogen and oxygen — and on Earth they are locked together to form vast quantities of water, an essential requirement of life as we know it. They are also the two most common elements in your body — it is a famous cliché, but nonetheless startling when you think about it, that 65 percent of the mass of your body is made up of water (most of it in the jellylike stuff that fills up the individual cells of your body). And if you leave the water out of the calculation, half of the remaining mass (the dry weight of your body) is carbon, 25 percent is oxygen, and just under 10 percent is nitrogen. Although the traces of other stuff in our bodies are very important for the processes of life, we are basically made out of carbon, hydrogen, oxygen, and nitrogen — the four most common reactive elements in the Universe. It doesn't necessarily follow that there is anything mystical about this — it need not be the case that the Universe was set up in some way to manufacture the kinds of things that life needs; rather, life can be seen as having evolved and adapted to make use of the raw materials that happen to be at hand. From this perspective, it's no more surprising that we are made out of carbon, hydrogen, oxygen, and nitrogen than that igloos are made out of blocks of ice or that houses in hotter climes are made from bricks of dried mud. In each case, the building blocks used come from the raw materials that are available.

These four elements (carbon, hydrogen, oxygen, and nitrogen) are so common in clouds of gas and dust in space (the kind of clouds that stars and planetary systems form from), and so regularly seen together, that they are sometimes referred to simply by the acronym CHON. Explaining where the hydrogen came from is no problem — it has been around ever since the Big Bang. So most of the puzzle of explaining the origin of the stuff we are made of boils down to explaining the origin of carbon, oxygen, and nitrogen — how they were manufactured out of primordial hydrogen and helium and scattered to form the clouds from which new stars formed.

To set the scene, I have raced ahead a bit, using terms like *atom* and *nucleus,* which are familiar to most people (if only in a vague sort of way), without bothering to spell out exactly what they mean. But maybe it is time to pause and take stock of what we are talking about. The story that follows — the story of CHON and of ourselves — is simple and straightforward, but it involves things that most people are brought up to believe are hard, like nuclear physics. But nuclear physics isn't hard, at least not the concepts — though working out equations and using them to construct computer simulations of what goes on in the heart of a star is. But once that work has been done, it is easy to understand nuclear physics in more general terms, without the mathematics. Perhaps "understand" is too strong a word, since some of the concepts run counter to common sense. But it is certainly possible to portray what is going on in words and pictures, without using equations.

The key to all this is the concept of a particle. What do we mean by the term *particle* in this context? Many people today are familiar with the idea of atoms as the ultimate building blocks of everyday matter, the smallest units of an element (a pure substance like oxygen or lead or aluminum) that can take part in chemical reactions, combining with other atoms. Remember that although the molecules are often a combination of atoms of different elements, creating compound molecules such as carbon dioxide, atoms sometimes combine with atoms of the same element, making molecules of things like oxygen, in which a couple of oxygen atoms are linked together. But one of the things that seldom comes across properly when we learn about atoms in school is just how tiny they really are. Atoms are only about 10^{-8} cm across, so it would take a hundred million atoms side by side to stretch along a line 1 cm long. But atoms themselves are not the smallest particles that are important to life, nor, as was thought until the 1890s, are they indivisible particles. Their chemical properties — the reasons why they link up in certain combinations but not in others — are explained in terms of the arrangement of much smaller particles, electrons, in the outer parts of the atoms. Electrons are almost unimaginably minuscule. The size of an electron compared to a speck of dust floating in the air is roughly proportionate to the size of that speck of dust compared to the size of the Earth. And yet, the properties of electrons determine the nature of all chemical interactions, including the chemistry of life. I do not intend to go into details of how chemistry works here (I covered that ground in my book *Almost Everyone's Guide to Science*); but the key chemical property of an atom is the number of electrons it possesses — and that number is itself determined by the next layer of structure within the atom, which is where nuclear physics comes into the story.

The nucleus is the core of an atom, where most of its mass is concentrated. In terms of its radius, the nucleus is a hundred thousand times smaller than the atom — 10^{-13} cm across compared to 10^{-8} cm across. It would take ten thousand billion average-sized nuclei to stretch along a line 1 cm long (to put this number in some sort of perspective, it is a hundred times the number of stars in the Milky Way Galaxy). The nucleus of hydrogen, the simplest kind of atom, consists of a single particle, called a proton. Each proton has one unit of positive electric charge, and each electron has one unit of negative electric charge. All atoms have the same number of electrons on the outside as they have protons in the middle and are therefore electrically neutral — in the case of hydrogen, one proton and one electron. The number of protons in the nucleus (called the atomic number) is what determines the number of electrons in the atom, and hence its chemical properties. The number of protons in the nucleus determines which element an atom belongs to — whether it is an atom of gold, hydrogen, silicon, or whatever.

In all atoms except hydrogen, there are both protons and particles called neutrons in the nucleus. The neutron is similar to the proton but has no electric charge. The masses of the proton and the neutron are almost the same — each, in round terms, is about two thousand times the mass of an electron, so it is really by far the bulk of the mass of an atom that is concentrated in its tiny nucleus.

But if the nucleus is packed full of particles with positive electric charge, why doesn't it blow itself apart? We all learned in school that positive charges repel one another, but (in my school at least) nobody ever bothered to explain why this basic rule of physics doesn't seem to apply inside the nucleus of an atom. When physicists were confronted with this problem in the 1930s, though, they were quick to see a way around it. Neutrons and protons (known collectively as nucleons) must be bound together in the nucleus by a strong force, which is called, logically enough, the "strong force." This force, which affects both kinds of nucleon, is about one hundred times stronger than the electric force, and the presence of neutrons in the nucleus helps it to bind protons together in spite of the natural electric repulsion between all the positively charged protons. But if you had more than about a hundred protons in a nucleus, the electrical repulsion would blow it apart in spite of the strong force, which is why there are no stable elements which have atoms containing more than about one hundred protons (and, of course, one hundred electrons). This is a delightful example of the way properties of tiny things like protons and neutrons affect the everyday world — the number of elements depends on the relative strengths of the strong force and the electric force. Although we don't notice the strong force in everyday life — because, unlike the familiar forces of gravity and electromagnetism, it has a very short range and can only make itself felt across a distance roughly the size of an atomic nucleus[2] — the variety of stuff in the world around us is direct testimony not only to its existence but to its strength.

Nuclear physics is all about protons and neutrons (nucleons), and we don't have to worry here about deeper structure within these entities, things like quarks. We need worry only about protons and neutrons and how they join together to make nuclei — even the chemistry follows automatically from the number of protons present in the

2. Which is why nuclei have the sizes they do.

nucleus, since there must be the same number of electrons to keep the atom electrically neutral overall. There are two key things about nuclei worth keeping in mind. The first is that hydrogen is a special case, because its nucleus consists of a single proton, with no neutron alongside it. Together with the fact that it has only a single electron to shield its positive charge from the attention of other atoms, this means that the positive charge on the nucleus of a hydrogen atom is less well hidden than the charge on the nucleus of any other atom and still has some potential to interact with other atoms, in spite of the presence of its lone electron. This is crucial to the story of life as we know it. The second important thing about nuclei is that the combination of two protons and two neutrons together in one nucleus makes an extremely stable unit — so stable that it was originally thought to be a single particle and is still known as an alpha particle. An atom which has the alpha particle as its nucleus, and therefore two electrons in the outer part of the atom, would be an atom of helium. So the alpha particle is also known as a helium nucleus (strictly speaking, a nucleus of helium-4, using an obvious labeling system in which the number indicates how many nucleons there are in the nucleus being referred to).

That's as far into the particle world as we have to delve to understand the origin of the elements, including the elements that make up the chemical compounds — the molecules — in living things like yourself. Protons, neutrons, and electrons can all be regarded as particles in this connection, and the special combination of two protons with two neutrons to make up a helium nucleus can also be regarded as a single particle for many purposes (there is one other entity which will come into the story, the neutrino, but that can wait its turn). Protons and neutrons in different combinations make up nuclei, and when nuclei are dressed in a cloak of electrons, they become the atoms of different elements. So how are the atoms of CHON arranged (with a smattering of other elements) to produce life as we know it?

The most interesting molecules in your body (and in all living things) are proteins. This may come as a surprise, in view of all the publicity that DNA, the carrier of the genetic code, has received in recent years. But although the message carried by DNA is important (nothing less than the blueprint, or recipe, which describes how to make and maintain a living organism), DNA molecules themselves are dull, dull, dull. It's a bit like a book—the concepts and ideas expressed in a book may be astonishing and amazing, but the string of letters representing those ideas is no more than the twenty-six letters of the alphabet (if the book is written in English), broken up by punctuation marks and arranged in a certain order. There is nothing intrinsically interesting in a jumble of letters; it is our conventional interpretation of what certain patterns of those letters (words) mean that makes books interesting.

DNA is even more boring than the English alphabet, because it uses an alphabet of just four letters to convey its message. Those four letters are represented by four chemical subunits called bases, strung out along the famous double-helix molecule of DNA and usually referred to by the initial letters of their chemical names, as C, G, A and T. This genetic code operates to store and convey information exactly like a four-letter alphabet, with three-letter words spelling out messages such as TCC CGG ACT GCT GCA and so on. This is certainly dull (if you don't know the code or speak DNA language) and looks like a restrictive form of communication when compared to the richness of the twenty-six letters of the English alphabet; but don't be fooled. Remember that computers use an even simpler "alphabet," a binary code that uses only two "letters," o and 1. A string of binary code might run 00101101110011010001110 and so on in a seemingly boring string of digits — but you can buy a CD ROM which contains the complete works of William Shakespeare, all stored as strings of 0s and 1s. Just as with DNA and the molecules of life, it is the message that is interesting, not the messenger.

And the message contained in DNA, in those three-letter words of DNA language, tells the cells of a living thing how to make proteins. Proteins are the most interesting and important molecules as far as the running of the cell is concerned (and therefore of any living thing, since all known living things are made of cells). Proteins provide both the structure of the cell and its machinery — both the factory and the workers. They determine what kind of cell it is, how it grows and divides, and how it uses energy to encourage the chemical reactions that make more molecules of life.

Proteins are wonderfully complicated molecules that come in a rich variety of shapes and sizes. There are structural proteins, such as the fibers which make up your hair or the hard covering of a cockroach. And there are worker proteins, such as the hemoglobin molecules that carry oxygen around in your blood or the insulin that enables the cells of your body to make use of the energy stored in glucose. There are more than fifty thousand varieties of protein in your body, and proteins are so much *the* molecules of life as we know it, as far as both structure and chemical activity are concerned, that almost right up until the time that James Watson and Francis Crick discovered the nature of DNA, at the beginning of the 1950s, DNA itself was regarded merely as some kind of scaffolding material that provided a structure for protein molecules to cling on to inside the cell but took no active role in the chemistry of life.

Protein molecules have such wonderful variety and versatility because they are made up of long chains of subunits called amino acids. On their own, amino acids have no particularly interesting properties — they are not alive in any sense. But strung together in the right way, they become complicated protein molecules which fold up upon themselves in specific ways, making protein molecules with very well defined shapes that interact with one another and with less complicated molecules to perform all the activities of life. Just twenty amino acids are used as the components of all proteins by all living things on Earth.[3] Although other amino acids exist, they are not used by life. And all living things use the same genetic code — the same three-letter words in the DNA code spell out the same message in all living cells. These are persuasive pieces of evidence to support the idea that all of life on Earth today is descended from a single life form (perhaps even a single cell) that first evolved the ability to make use of DNA and proteins in this way. All the fifty thousand plus proteins in your body, and all the proteins in all other living things we know of, are made out of the same twenty amino acids, just as all the words of Shakespeare, and all the other words written in the English language, are made out of the same twenty-six letters of the alphabet.

DNA comes back into the story because what those three-letter words in the DNA alphabet actually code for is amino acids. In DNA language, the "word" GAC, for example, means "make a molecule of aspartic acid," whereas the word AGG translates as an instruction to the machinery of the cell to make a molecule of the amino acid arginine. You can also think of protein molecules, made of strings of amino acids arranged in a certain order, as messages written in a language. It is the variety of amino acids that makes protein molecules so interesting.

3. There are two other amino acids that are found in a very few proteins, and one of the twenty I have mentioned comes in two marginally different forms; so you may see elsewhere references to either twenty, twenty-one, or twenty-three amino acid building blocks for proteins. But twenty is a nice round number, and I shall stick with it.

The amino acids. Note the importance of CHON in the chemistry of life.

Amino acids get their name because their structure includes a group of atoms formed from a molecule of ammonia. An ammonia molecule consists of a single atom of nitrogen linked by chemical bonds to three atoms of hydrogen. If one of these hydrogen atoms is replaced by something else, the residue of the ammonia molecule is called an amine group. In amino acids, the amine group is attached to a carbon atom, which is itself linked to other atoms. One of the other things the carbon atom is always attached to in an amino acid is a group of atoms called the carboxylic acid group, which is where the other part of the name comes from. This is a combination of one more carbon atom, two oxygen atoms, and one hydrogen atom, which can be written as -COOH (the dash denotes the bond which, in this case, attaches to another carbon atom that is itself attached to the amine group). The carbon atom to which both the amine group and the carboxylic acid group are attached still has the capacity to make two other chemical bonds, linking up with other atoms and groups of atoms, providing the variety of amino acids found in living things. Those other groups are themselves made up almost entirely of carbon, hydrogen, nitrogen, and oxygen, with just the odd atom of sulfur. CHON is right at the heart of the structure of proteins.

DNA molecules are also made almost entirely out of CHON. The four chemical subunits that make up the letters of the DNA alphabet (the ones known as bases), indeed, contain nothing else; but the spine of the DNA molecule, which holds the bases together in a long sequence spelling out a message in the genetic code, is held together by a combination of atoms called the phosphate group, and each of these, as the name implies, contains one atom of phosphorus. Each phosphate group is linked to a molecule of a sugar known as deoxyribose, each of which has a base sticking out from its side. The sugar molecule is linked to another phosphate group, which is linked to another sugar molecule (with base attached), and so on.

It's worth digressing just a little to comment on two subtle features of the chemistry of life, features of the properties of two of the components of CHON that help provide the richness and variety of the living world—such essential features that some people have argued that there may, in fact, be more to the story of life than simply making

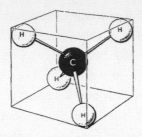

The structure of the simple hydrocarbon methane indicates the way carbon atoms can form four widely separated chemical bonds with other atoms — a key feature in the chemistry of life.

use of the building blocks that happen to be at hand. The first, rather obvious, feature of the chemistry of life is that it is built around carbon — so much so that the study of carbon chemistry also goes by the name organic chemistry. Carbon is so important in the chemistry of life because an atom of carbon has the ability to form chemical links with four other atoms (perhaps including other carbon atoms) at the same time. Hydrogen, for example, can make only one regular chemical bond with another atom; oxygen has the ability to make two. Chemical bonds are links between atoms formed by the electrons in the outer parts of those atoms — in effect, one electron from each atom comes under the influence of both nuclei, so that the pair of electrons forms a bond holding the two atoms together. For reasons connected to the way electrons are arranged in the outer parts of atoms, no atom can make more than four bonds at once, and carbon is the best atom of all at doing this. Because one of those bonds might be with another carbon atom, it is possible for carbon atoms to form long chains, with yet other interesting things attached to the sides of the chains, or rings, with interesting things attached around the edge of the ring as well. This is what gives carbon chemistry — organic chemistry — its richness. There are other atoms which can make four bonds at a time — silicon, for example — but, as we have seen, they are much less common than carbon. So it may be the combination of its availability and its chemical bonding ability that makes carbon so important to life.

The other intriguing property concerns hydrogen, the simplest atom. I was careful to say that it can form only one regular chemical bond at a time, because it can also form a weaker kind of link with another atom, because of the way in which its lone proton is imperfectly shielded from the outside world by its single electron. Especially when that electron is used in making a chemical bond with another atom (for example, in a molecule of water, where two hydrogen atoms are attached to a single oxygen atom), the positive charge on the proton can still influence, and be influenced by, the negative charge on electrons on other nearby atoms. This has the effect of making some molecules that include hydrogen atoms slightly sticky, in an electrical sense. When water molecules brush past one another, the hydrogen atoms in one molecule are attracted to the oxygen atoms in other molecules. It is this stickiness that makes water a liquid at the temperatures common on Earth today. By and large, the temperature at which a gas liquefies depends on the mass of its molecules, with heavier molecules condensing to form liquids at higher temperatures. But the molecular weight of a water molecule is just 18 units, on the usual scale, and something like carbon dioxide, with a molecular weight of 48 units (more than twice that of water), is a gas at room temperature, while water is a liquid.

This affinity of hydrogen atoms that are already attached to one molecule for other nearby molecules is called the hydrogen bond. It is weaker than the usual form of chemical bonding, but none the less real, and very important for life as we know it. The two strands of the double helix of DNA are held to one another entirely by hydrogen

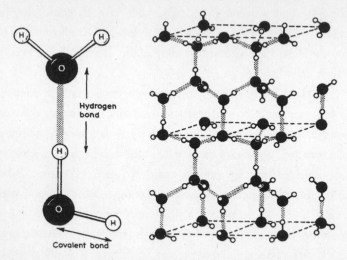

Because of the special properties of hydrogen atoms (discussed in the text), they can form a relatively weak bond, called a hydrogen bond, with some other atoms in appropriate circumstances. Here, the hydrogen bonds operate between water molecules, making water liquid at relatively high temperatures and encouraging ice to form a very open solid structure.

bonds, and in a very specific way. It happens that when the two bases known as T and A are lined up opposite each other in the right way (each attached, of course, to its own sugar molecule, which is part of the spine of a single strand of DNA), they can form a pair of hydrogen bonds loosely holding the two units together, rather like a two-pin plug fitting into a two-pin socket. The C and G bases link up in a similar way, but this time forming three hydrogen bonds, like a three-pin plug fitting into a three-pin socket. The pairings are quite specific: T and A pair with each other, but neither of them with either C or G; C and G pair with each other, but neither of them with T or A. This is why a DNA double helix can be "unzipped" by the machinery of the cell, and each strand used as a template on which that cell

Two pairs of bases in molecules of DNA held together by hydrogen bonds. This is a key feature of life as we know it; once again, notice the importance of CHON.

machinery (in the form of protein molecules) can build a replica of the missing strand, so that you have two identical DNA molecules where there used to be one — an essential step in reproduction.

Keep in mind that there is something more to CHON than the simple fact that it is the most available stuff around for use in the chemistry of life. There is, though, still another aspect to the story. In order for life to exist, it has to have a supply of chemical raw materials and energy — and the only place where we know for certain that life exists is on our home planet, Earth. Is there anything special about the Earth that makes it a suitable home for life?

Because we are considering only "life as we know it" in this book, it might seem that the conditions required for life to exist would be rare. Certainly, from one perspective the Earth is an unusual planet, even by the standards of our own Solar System, and we have no means (yet) of knowing whether our Solar System is typical of the kind of planetary systems that are associated with stars like the Sun, although

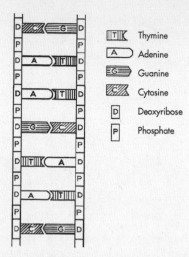

The many hydrogen bonds linking two strands of DNA create a ladderlike structure, the two strands of DNA running side by side.

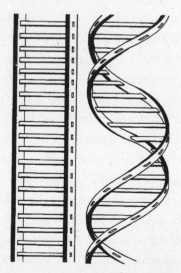

In fact, the "ladder" is twisted to make the famous double helix.

we do know there are many stars like the Sun. This kind of argument is sometimes used to suggest that life as we know it is doubly rare — that it requires an unusual planet that is itself part of an unusual planetary system. I do not find the argument persuasive, because a closer look at the Solar System suggests that it may, in fact, be unusual in the opposite sense — the evidence suggests that it might be rare to find life on one planet in a planetary system not because most planetary system have no homes suitable for life, but because most planetary systems have more than one home suitable for life.

The key to the existence of life as we know it is water. So much so that the word we use for a region without liquid water is the same as the word we use for a region without any life — desert. There are sound reasons for this, based on the physical and chemical properties of water. Water is very good at holding other things (other chemicals) in solution, allowing those other things an opportunity to interact with one another. It also protects them from some of the harsher aspects of the outside environment, such as damaging ultraviolet radiation, which would have scoured the surface of the Earth before an oxygen-rich atmosphere developed. And, in the right range of temperatures (a surprisingly large range of temperatures, common on Earth), water doesn't just sit around as solid ice, or slop around as a liquid, or float in the air as a gas — it does all three things at once, so that all three phases of water (solid, liquid, and gas) coexist in a dynamic equilibrium, with molecules constantly being swapped between the three phases. This helps to spread water round the planet: moisture evaporates from the sea to become a gas, and it falls as rain to make rivers and runs back to the sea. And this enables life to exist on land as well as in the oceans (indeed, it is an absolute prerequisite for the existence of life forms like us on the land surface of the planet). So, for the purpose of the present discussion, I am going to place an even more restrictive (but simple) requirement on the existence of life as we know it. I am going to consider a planet as a potential home for life only if it has liquid water.

These important properties of water are partly due to its ability to form hydrogen bonds, so they are intimately connected with the fundamental properties of atoms. The hydrogen bond is also important in another unusual property of water that is certainly important for the existence of life on Earth, and may be of more universal significance. Ice floats. This is such a familiar phenomenon that the immediate reaction is, so what? But think about it. You would be amazed if you saw a puddle of liquid lead, say, with a lump of solid lead floating on top of it, no matter how carefully the temperature of the liquid was being manipulated. In the normal run of things, solids are more dense than their own liquid form, so a solid lump of anything ought to sink in a liquid puddle of that same stuff. Water does not do this solely because of the way hydrogen bonds form between the molecules. Because individual molecules have a distinctive V shape (with the oxygen atom at the point of the V and the two hydrogen atoms at the ends of its arms), when water solidifies, the hydrogen bonds between hydrogen atoms in one molecule and oxygen atoms in other molecules hold the arrangement of molecules together in a very open lattice (actually very similar to the lattice in a crystal of diamond, though not so strong; see figure on p. 30). As a result, the water molecules are actually slightly farther apart than they are in the liquid form when it is just above freezing. In the liquid, the molecules brush past one another, without the hydrogen bonds "setting," as it were. But when the temperature drops to freezing, the molecules snap into place in the crystal lattice.

So ice floats on water. Why is this important to life? Not least because the ice floating on top of a cold ocean acts as a lid that holds any remaining heat in, stopping evaporation from the surface, which would cause the ocean to cool still more. There have been many Ice Ages on Earth, occasions when large parts of the ocean surface (like the Arctic today) were covered in floating ice. If ice behaved like any self-respecting solid and sank in its own liquid, when an Ice Age occurred, the oceans would freeze from the bottom up. Ice would form at the surface as the water cooled and then sink to the bottom, more ice would form at the surface and sink to the bottom, and so on. The process could continue until the oceans were one solid lump of ice, with no liquid water left — and if that ever happened, it would be extremely difficult to melt them again, because even if the climate changed, the ice ball would reflect away a lot of the incoming heat from the Sun, and thawing the ice would require a huge input of energy.

Exactly the same effect can be found in a garden fishpond. The layer of ice floating on the pond in winter keeps the water underneath liquid and maintains a suitable environment for the goldfish. If the pond froze from the bottom up, their fate would be sealed by even a relatively modest cold snap. In a severe freeze, the right thing to do for the fish is to maintain a small hole in the ice, so that oxygen gets into the water; the worst thing to do is to break up the ice layer completely, which only exposes more liquid water to the air and encourages it to freeze.

All this is particularly relevant, and I have gone into the story in such detail, because we do now know of a potential home for life in the Solar System that exists simply because ice floats on water. One of the great surprises of the age of space exploration came in the late 1990s, when the probe *Galileo* visited the Jupiter system and made repeated passes by the moon Europa, sending back pictures and other data. The evidence overwhelmingly suggests that much of the surface of Europa (perhaps all of it) is covered by floating pack ice, very similar to the kind of ice we see in the Arctic. Europa is just (only just!) warm enough for liquid water to exist there, because the tidal forces of Jupiter, constantly squeezing the rocky core of the moon in and out, generate heat in its interior. Without that covering of ice, though, the heat would quickly escape and the moon would freeze. Thanks to that skin of ice, Europa is now regarded as a potential home for life, and there are plans for future space missions that will drop probes through the ice to investigate the water below.

There is another potential life-bearing moon in the outer part of the Solar System. Titan, the largest moon of Saturn, with a diameter of 5,150 km (50 percent bigger than the diameter of our Moon), has a thick but cold ($-180°C$) nitrogen atmosphere. It has been described as resembling the early Earth, before the advent of life, put into deep freeze. If something happened to warm Titan up to the point where liquid water could flow, it is possible (perhaps even probable) that the processes which led to the evolution of life on Earth would occur there.

This is relevant because of another surprise that confronted astronomers in the late 1990s. As I have mentioned, there is now evidence of large planets in orbit around about a dozen stars, as revealed by the way the planets tug on their parent stars gravitationally, jiggling them to and fro. As yet, only large planets can be detected in this way, because the jiggling effect is so small that it cannot be detected for Earth-sized planets. The big surprise is that most of the giant planets so far detected in this way turn out to be very close to their parent stars, in some cases even closer than the Earth is to the Sun (for comparison, Jupiter, the largest planet in our Solar System, is five times farther from the Sun than we are). This is usually interpreted as bad news for the possibility of finding life in these planetary systems, because a giant planet in such close orbit around its parent star will exert a gravitational influence on any Earth-like planets in Earth-like orbits that will eject them entirely from the planetary system, in a kind of gravitational slingshot effect. But that argument misses the point that if such a planet has a family of moons, and if some of those moons are similar to Europa or Titan, then there is every chance that conditions on those moons will be right for life. Because Jupiter has four large moons and Saturn another four (taking "large" to mean anything with a diameter bigger than a thousand kilometers), if anything the existence of a giant planet in an Earth-like orbit increases the number of potential homes for life in a planetary system—quite apart from the possibility of life forms rather different from those we know on Earth existing in the atmospheres of the giant planets themselves.

So we know of at least two "near misses" in the outer regions of our Solar System, beyond the asteroid belt that marks the limit of the orbits of the small, rocky planets. But even if we restrict our attention to those four inner planets (sometimes called the "terrestrial" planets, to distinguish them from the gas giants, or "jovian" planets), we find that it may be a case of bad luck that there are not at least two inhabited planets orbiting our Sun, rather than it being good luck to have even one.

We can forget about Mercury, the planet closest to the Sun. It is an airless desert world, like our Moon, with a battered surface, also rather like our Moon. With a diameter of 4,880 km, Mercury is intermediate in size between the Moon and Mars, but with no atmosphere to smooth out temperature differences by blowing winds around the planet, temperatures at the surface of Mercury range from several hundred degrees Celsius at local "noon" to −180°C at midnight. This is definitely no place for life as we know it.

At first sight Venus, the next planet out from the Sun, scarcely looks any more prepossessing, even though it has 82 percent of the Earth's mass and a diameter of 12,104 km, thereby resembling the Earth (which has a diameter of 12,756 km) most closely in terms of size and distance. Whereas Mercury has no atmosphere, as far as life forms like us are concerned, Venus has too much. Its thick atmosphere, made mostly of carbon dioxide, exerts a surface pressure that is ninety times greater than the atmospheric pressure on Earth at sea level. Because Venus is closer to the Sun than we are, you would expect it to be hotter than the Earth. But the greenhouse effect of that thick carbon dioxide atmosphere has taken its temperature to extremes, the surface temperature on Venus reaching, at least, a searing 450°C. The planet is almost entirely covered in high-level clouds, which reflect light from the Sun and make it a bright object in the sky, often visible in the early morning or early evening — but those clouds are laced with sulfuric acid, so what rain does fall toward the surface of Venus (actually evaporating long before it reaches the surface) is highly acidic.

When you compare Venus with Earth, though, and look at the geologic evidence for the way the Earth obtained its much more modest blanket of atmosphere, an intriguing picture emerges. If you first compare the Earth with the Moon, you can see the power of the greenhouse effect. The average temperature on the Moon (taking into account all latitudes and both night and day) is −18°C. This measured temperature exactly matches the temperature that such an airless ball of rock in space, at the distance the Earth and the Moon are from the Sun, ought to have, according to the known laws of physics describing the way things like lumps of rock absorb heat and reradiate it. Among other things, this means that the temperature of the Earth, if it had no atmosphere or oceans and was just a ball of rock in space, would also be −18°C. In fact, the average temperature over the surface of our planet is close to 15°C.

The difference (some 33°C of warming) is entirely due to the greenhouse effect of gases in the atmosphere of the Earth, particularly carbon dioxide and water vapor. Energy from the Sun, in the form of sunlight, passes through the atmosphere almost unimpeded and warms the surface of the planet. The warmed surface reradiates energy, but at longer wavelengths, in the infrared part of the spectrum. Some of this infrared radiation is absorbed in the lower part of the atmosphere and keeps the surface of the Earth warmer than it would otherwise be. This is what is known as the atmospheric greenhouse effect—and, again, the increase in temperature observed here on Earth exactly matches the calculated warming that ought to be produced by the measured amount of carbon dioxide and water vapor in the air. From this perspective, the greenhouse effect is a good thing—without it, we would not be here. Climatologists (and even politicians) are, however, currently concerned about the greenhouse effect because human activities (the burning of fossil fuel and the like) are adding to the amount of carbon dioxide in the air, strengthening the greenhouse effect and making the world warmer still (an increase sometimes referred to as the anthropogenic greenhouse effect), with potentially damaging consequences for agriculture.

The carbon dioxide that occurs naturally in the Earth's atmosphere today is only a minor component, which is why human activities can have a proportionately large effect. The gas came originally from volcanic activity, known as "outgassing," which released carbon dioxide, among other things, from within the interior of the planet. Equivalent processes created the atmospheres of Venus and Mars. Over geological time, it is estimated that the amount of carbon dioxide released in this way from inside the Earth was equivalent to sixty to seventy times the amount of gas in our present-day atmosphere — not just sixty to seventy times the amount of carbon dioxide but sixty to seventy times as much as everything in the present-day atmosphere put together, in the sense that it would provide a pressure at the surface of the Earth today sixty to seventy times greater than the present atmosphere does. If all that carbon dioxide had stayed in the atmosphere, there would have been a runaway greenhouse effect producing conditions very like those we see on the surface of Venus today. Instead of Venus being a twin of the Earth, Earth would be a twin of Venus. So why didn't it happen?

By the time the temperature on either Venus or Earth rose to the boiling point of water, all the water would have turned to vapor, increasing the strength of the atmospheric greenhouse effect still further. The reason why Earth did not end up like Venus is that from its very beginning, as soon as it had formed an atmosphere, there was liquid water on Earth. Carbon dioxide dissolved in the water and was laid down in the form of carbonate rocks — it is by measuring the amount of carbonate laid down in rocks that geologists know how much carbon dioxide has been outgassed in the life of the Earth so far. Because Venus is just that crucial bit closer to the Sun than we are, from the early days of the Solar System it was just too warm for oceans of liquid water to form there. With no oceans, all the carbon dioxide being outgassed has indeed added to the water vapor in the atmosphere and built up to become the thick atmosphere we see today, with a strong greenhouse effect. And it is no coincidence that the atmospheric pressure on Venus today, ninety times that on Earth, is roughly the same as the pressure our atmosphere would exert if none of the carbon dioxide had been laid down in rocks. Similar planets have outgassed similar amounts of carbon dioxide. If the Sun had been a little cooler, or if Venus had been in an orbit a little farther out from the Sun, Venus might really have become a twin of the Earth, with oceans, a thin atmosphere, and life.

Now look a little farther out from the Sun than we are. There we find the planet Mars, which has a very thin atmosphere consisting almost entirely of carbon dioxide. It is a red desert today, with no signs of life. But that desert is marked by clear evidence that water once flowed there, notably the huge canyons, the river valleys, and other geologic features that are strikingly like those formed on Earth through the activity of subsurface water. It seems that when Mars was young it had a reasonably thick atmosphere with a greenhouse effect strong enough to allow water to flow. But because Mars is much smaller than the Earth (measuring only half the diameter of the Earth and with just one-tenth of its mass), its interior cooled very quickly and its geologic activity stopped long ago, so that no more carbon dioxide was outgassed. And, again because the planet is so small, the atmosphere that had built up by then was gradually lost into space, because the gravitational pull of Mars was inadequate to hold on to it. As the atmosphere thinned, the greenhouse effect weakened, and the planet froze—although there is almost certainly a large amount of water still there, frozen below the surface. If Mars had been as big as the Earth, we could well have had a twin planet on that side of our orbit around the Sun.

Or look at it in terms of the temperature of the Sun. After all, not all stars have the same brightness. You could say (some people do) that we are lucky that the heat of the Sun is just enough to make the Earth a suitable home for life. Not too warm and not too cold, but, like Baby Bear's porridge, "just right" (this is sometimes called the Goldilocks Effect). But if the Sun had been a little cooler, Earth and Venus would have both been potential homes for life, and if it had been yet a little cooler there would still have been water on Venus, even if the Earth froze. And if the Sun had been a little warmer, even if Earth had followed Venus into the runaway greenhouse trap, Mars could have survived as a pleasant planet long enough for life to evolve. It seems as if, for a planetary system like our own, it is very hard to avoid having at least one watery planet, as the bare minimum. That is why my own view is that we are rather unlucky not to have at least one neighboring planet, within comfortable reach of our existing space-faring ability, that is already suitable for life as we know it. All evidence points to the likelihood that other planets (or moons) just right for life as we know it will exist in orbit around at least some of the stars we see in the sky. So let's step out into the Universe at large. We don't have to worry about exactly how planets form or exactly how life originates. We know that happens if you have a supply of CHON and a planet like the Earth. But can we explain where the stuff we are made of, dominated by CHON, comes from? That means understanding the way the stars work — starting with a star like the Sun.

3

The Stars Are Suns

As with the meaning of terms such as *atom* and *nucleus*, I have assumed so far that anyone reading this book is likely to be aware that the Sun is a star and that the essential reason why it appears so big and bright in the sky, compared with the pinpoints of light that are the other stars, is that it is so much closer to us than they are. But this wasn't always obvious. Indeed, in antiquity the stars were literally regarded as pinpoints of light—tiny holes in a spherical shell of dark material surrounding the Earth, through which we could see light shining in from outside. This was not a completely crazy idea at the time, for two reasons. First, the stars seem to be fixed in the same places relative to one another in the sky, forming the patterns known as constellations, so that it was reasonable to think they might all be attached to some structure rotating around the Earth. Second, there weren't many stars to account for in this way. Over the entire sky, only about six thousand stars can be picked out with the naked human eye, even if there is no artificial light (or moonlight) to dazzle you. You might think this would mean that there are three thousand stars that can be seen at any moment during any night, since only half the sky is visible at any one time. But faint stars low on the horizon tend to get lost in the haze, and hills and trees block some of the view, so that more realistically, perhaps only a couple of thousand stars, at most, are visible at any one time. These are sensible numbers that fit in with the human scale of things and that helped make stars seem like something within a human framework. More or less this picture held sway for thousands of years; things changed only at the beginning of the seventeenth century, when Galileo Galilei turned his telescope on to the Milky Way and discovered that it is made up of myriad individual stars, which appear to the naked eye as a single white cloud.

What's important here is that from the very beginning of the scientific investigation of stars, technology comes into the story. Progress in astrophysics, the study of the stars, has been possible only with the help of technology, for no matter how clever you are or how good your theoretical ideas may be, if you have no way of testing those ideas by comparing them with observations, there is no hope of finding out which ideas are correct and worth exploring further and which are wrong and must be discarded. This is why the ancients suggested that phenomena they could see in the skies might be the work of gods — literally, heavenly phenomena. Among the many such stories about the origin of the Milky Way, the one that has given it its modern name comes from Greek mythology, which portrayed the white ribbon of light across the sky as milk spilled from the breast of the goddess Juno while feeding the infant Hercules. The ancient Greeks (and their counterparts in other cultures) came up with such ideas not because they were any less clever than modern astronomers but because they had far less information to work with. Just how clever some of the ancients were is highlighted by the speculations of the Greek philosopher Democritus, who lived in the fifth century B.C. (some two thousand years before Galileo) and suggested that the Milky Way might indeed be made up of countless stars that were too faint to be picked out individually but that together blended into a glowing band across the sky.

Democritus was also a leading early proponent of the atomic theory (or atomic hypothesis, as it was at that time — strictly speaking a theory is a hypothesis that has been tested against experiment and observation and has passed those tests). But in both cases, he had no way to test his ideas, because he lacked the appropriate technology. They remained hypotheses, not theories, until the technology to test them was invented. If Democritus had lived and worked in the second half of the sixteenth century, though, he might well have made as profound an impact on the development of science as Galileo, who wrote in his book *The Starry Messenger* (published in 1610):

I have observed the nature and the material of the Milky Way. With the aid of the telescope this has been scrutinized so directly and with such ocular certainty that all the disputes which have vexed philosophers through so many ages have been resolved, and we are at last freed from wordy debates about it. The Galaxy ["galaxy" is Greek for "milky way"] is, in fact, nothing but a congeries of innumerable stars grouped together in clusters. Upon whatever part of it the telescope is directed, a vast crowd of stars is immediately presented to view. Many of them are rather large and quite bright, while the number of smaller ones is quite beyond calculation.

With modern technology — modern telescopes — we can go one better than Galileo. By counting the number of stars seen in a small patch of the Milky Way and multiplying to cover the area of the entire Milky Way, astronomers now estimate that our Galaxy contains around two hundred billion stars, a number quite beyond the realm of everyday human experience.

About a hundred years after Galileo made these discoveries, the other leg of the commonsense idea of stars as tiny lights attached to a rigid sphere surrounding the Earth was finally knocked away. Edmond Halley, of comet fame, was appointed by the Royal Society to collate a new star catalog, using data from observations which had been carried out by the first Astronomer Royal, John Flamsteed.[1] In the course of this work, Halley compared data from a catalog compiled by Hipparchus in the second century B.C. with the new data. Of course, many more stars had been catalogued by Flamsteed, but his catalog included the bright stars studied by Hipparchus. Halley found that in most cases the data from the two catalogs matched, showing that the ancient Greeks had been skilled observers who had accurately measured the positions of the stars in the sky. But in a few cases there were striking differences between the star positions given by Hipparchus and the positions of those same stars observed in the eighteenth century. The conclusion was inescapable: some of the stars had moved across the sky in the intervening centuries. They were not fixed to a single framework at all but could move independently of one another.

By then, though, it was already being suggested that the stars were Suns. Between the time of Galileo and the time of Halley, several

---

1. This was a matter of bitter dispute at the time, since Flamsteed strongly objected to someone else using his data, but the wrangle that resulted need not worry us here.

astronomers had tried to estimate distances to the stars, by guessing that all stars are the same brightness as the Sun and look faint only because of their distance from us. Among other things, this would mean that fainter stars must be farther away than brighter stars, so they could not all be attached to the same crystal sphere around the Earth — but the crystal sphere idea had been well shattered by Galileo's discoveries. One of the people who tried to use this technique was Isaac Newton, who calculated that if the bright star Sirius was actually the same brightness as our Sun, it must be about a million times farther away from us than the Sun is. This is only about twice the modern estimate of the distance to Sirius and gives a genuine feel for the kind of distances you would have to travel to reach the *nearer* stars. Putting this in a slightly different perspective, Sirius is so far away from us that light takes 8.7 years to travel from Sirius to the Earth (so Sirius is 8.7 light years away); it takes just 8.3 minutes for light to travel from the Sun to the Earth (so we are 8.3 light minutes away from the Sun). And light travels at a speed of 300,000 kilometers per second.

Astronomers only got to grips with the true distances to the stars in the 1830s and 1840s, when they were (just) able to measure a few of these distances directly, using the geometrical technique of parallax. The nearest stars are close enough to us that they seem to shift across the sky by a tiny amount, relative to the more distant background of fixed stars, as the Earth moves in its orbit around the Sun. The position of a particular star against the background constellations is measured accurately at six-month intervals, when the Earth is on opposite sides of its orbit around the Sun. The apparent shift in position of the star being studied is called parallax and is measured in seconds of arc. For the nearest stars, the measured shift, or parallax, is typically a few tenths of a second of arc, corresponding to distances of up to ten light years. And to put the skill of the astronomers who first measured these parallaxes in perspective, the angular size of the full Moon on the night sky is about 30 minutes of arc. The largest measured stellar parallaxes correspond to apparent stellar displacements across distances roughly one sixtieth of 1 percent of the size of the full Moon in the sky.

Even by 1900, parallaxes had been measured for just sixty stars. If this had been the only way astronomers could measure stellar distances directly, they would still have had little idea about the nature of the different kinds of stars we see and how they work. But there is one other geometrical technique for measuring stellar distances which takes astronomers just far enough out into the Universe to begin to be able to understand the different kinds of stars. It works for clusters of stars that are moving together through space, like a shoal of fish swimming in the same direction through the sea. But it works only for clusters that are close enough for their motion to be detected, which involves taking photographs of the cluster at intervals years (or decades) apart and comparing the positions of the stars in the two photographs.

A group of stars moving in the same direction is in effect running along parallel lines, just as all the cars on an eight-lane superhighway are moving in the same direction. From off to one side, the lanes of the superhighway seem to converge on a point in the distance, the point where all the cars are heading. In the same way, measuring the movement of the stars in a cluster across the sky (over many years!) reveals that they all seem to be heading toward a particular point in the sky (a different point for each cluster, of course). This tells you where the stars in the cluster are heading.

At the same time, astronomers can measure the speed of the stars through space. It is easy to measure how fast they are moving across the line of sight (if you are patient enough or have very old records of where the stars used to be). This angular movement would tell you the distance to the star, if you knew its actual velocity through space. The faster it is moving, the farther it will shift through space in a single decade or a single century; but the farther away it is, the less movement will be perceived across the sky. But, of course, the star may also be moving toward or away from us, along the line of sight. Happily, it is easy to measure this component of velocity, using the Doppler effect, which became a useful astronomical tool in the second half of the nineteenth century. Predicted in 1842 by Christian Doppler, the Doppler effect is a change in the wavelength of light caused by the source of the light moving toward or away from you through space. Objects that are moving away stretch out the light to longer wavelengths (like stretching a spring), and because red light has longer wavelengths than blue light, this is called a redshift. Objects that are moving toward you squash up the wavelengths of light they are emitting (like squashing a spring) producing a blueshift. By measuring the amount of redshift or blueshift in the light from a star, you can measure the speed of the star (its real speed, through space) along the line of sight. In each case, there will be only one true speed through space across the line of sight that combines with this measured speed along the line of sight to make the overall motion of the star line up with the motion of the whole cluster toward the known point in the sky where the cluster is going. The known true speed of a star across the line of sight, calculated in this way, can then be compared with the angular rate at which we see it moving from Earth to determine how far away it must be to produce the observed angular motion across the line of sight.

I've explained this in some detail, because it is an absolutely crucial step in everything that follows. You don't have to be able to do the calculations yourself, but you do have to be convinced that astronomers really do have a good method of measuring distances to nearby clusters of stars — or at least to one particular cluster. The moving-cluster method gives a good distance to that one particular cluster, called the Hyades, which lies in the constellation Taurus. The Hyades contains more than 200 stars, spread over a small volume of space about 150 light years away from us. They are so far away that the distance from one side of the cluster to the other doesn't make much difference — roughly speaking, all the stars in the Hyades can be regarded as being at the same distance from us. But their degrees of brightness vary greatly. The stars must, therefore, really have different brightnesses — this is not an illusion caused by the fact that some of the stars are closer to us than others. And since you know the distances to each of the stars and their apparent brightnesses, you can work out their true brightnesses (or intrinsic luminosities), compared with that of the Sun. It is this kind of information that tells us that the Sun is about in the middle range of stellar brightness and is, in that sense, an average star (in another sense, it is brighter than average, since there are a lot more faint stars than there are bright stars). Even this fundamental piece of information has been known for sure only since the second half of the nineteenth century — scarcely more than a hundred years.

Distances to other clusters of stars, far beyond the Hyades, are determined in a variety of ways. Once certain kinds of stars, distinguished by their color, are known to all have a specific intrinsic brightness, or luminosity, they can themselves be used as distance indicators ("standard candles"), by comparing their apparent brightnesses with the expected intrinsic luminosity. The best distance indicators of this kind are stars which vary in a particular distinctive way (making then easy to pick out) and which all have roughly the same brightness (making them good standard candles). If you find one of these RR Lyrae stars, as they are known, in a cluster, then you can use it to calculate a distance to the cluster, before going on to compare all the individual stars in the cluster with one another to see how their intrinsic luminosities differ from one another. Then, of course, you can compare one whole cluster with other clusters, to learn yet more about the similarities and differences between stars.

Another kind of variable star, called Cepheids, is useful in measuring distances to other galaxies, beyond our Milky Way. But there is no need to go beyond the Milky Way just yet. The kind of distance determinations I have described are sufficient to show that the Milky Way as a whole is a disk-shaped system about 100,000 light years across and only 1,000 or so light years thick, containing an estimated 200 billion stars (though this started to become clear only in the 1920s, little more than a single human lifetime ago). The Sun sits in the disk, about two-thirds of the way out from the center of the Galaxy, orbiting around the center of the Galaxy more or less the way that the planets orbit the Sun.

The number of stars in the Galaxy is roughly the same as the number of rice grains that could be packed into a cathedral; but if the rice grains were spread out to make a scale model of the Milky Way in the right proportions, the model would have a diameter of about 400,000 km, the same as the distance from the Earth to the Moon.

Distances tell us the intrinsic luminosities of stars, and this is half the battle in determining how stars work. The other half of the story is their masses. There is actually only one way to determine the masses of stars precisely, and that is to observe two stars in a binary system orbiting around each other, the way the Earth and the Moon orbit around one another. Although, following Galileo's development of the astronomical telescope, astronomers had previously noticed pairs of stars that lie close to each other in the sky, it was only in 1767 that the British polymath John Michell (who, incidentally, came up with the idea of black holes) suggested that some of these pairs of stars might be physically associated, not just juxtaposed by chance, with one star relatively close to us and one much farther away along nearly the same line of sight, in the same way that at a certain time of night the Moon might be seen "next to" a particular constellation of stars. The first systematic observations of such double stars were carried out in the last quarter of the eighteenth century by William Herschel; his observations of the way some of these pairs of stars shifted around each other noticeably over the course of twenty years or so provided the evidence that these systems must, in his words, be "real binary combinations of two stars, intimately held together by the bonds of mutual attraction." In the nineteenth century, the study of binary stars became a key topic in astronomy, precisely because it is possible to determine the masses of the stars by getting the details of their orbits around one another. From the study of the orbits of planets in the Solar System, and from Newton's law of gravity and his laws of motion (themselves only published, in his great work *Principia,* in 1687), astronomers know that two simple equations describe the orbits of binary systems. One relates the distance between the two members of the binary to their combined mass (the mass of star A *plus* the mass of star B). The other equation relates the distance of each star from the center of mass of the binary system (its point of balance, if you like) to the ratio of the masses of the two components (the mass of star A *divided by* the mass of star B). And once you know both the total mass and the ratio of the two masses, it takes but a moment to work out the actual masses of the two stars.

Of course, it isn't quite that simple in practice — it never is in astronomy. You have to study the binary pairs for years, or even decades, to get accurate details of the orbits, and you have to make allowance for the way the orbit is oriented in the sky (whether we see it sideways, face on, or somewhere in between). You can get some idea of how painfully slow progress was from the fact that even in 1924, when the pioneering astrophysicist Arthur Eddington put together all the available information and plotted it on a graph relating the luminosity of a star (the absolute luminosity, for stars with known distances from us) to its mass, he had only a few dozen accurately determined stellar masses to work with. But these were enough to show two things: that there is a range of stellar masses from about one-fifth up to about twenty-five times the mass of the Sun; and (most important of all) that the mass of a star is related to its luminosity in a simple way.

In general, more massive stars are brighter than less massive stars. More specifically, for stars very like the Sun (with masses from about 0.3 to 7 times the mass of the Sun) the absolute luminosity is proportional to the fourth power of the mass (M^4), so doubling the mass of a star makes it sixteen times brighter; while for more massive stars the absolute luminosity goes as the cube of the mass (M^3), so doubling the mass of a star makes it "only" eight times brighter. As we shall see, this simple mass-luminosity relationship provides a valuable insight into the way stars keep shining by generating heat in their interiors. But even these first scientific discoveries about the workings of the stars were being made as recently as the mid-1920s. Astrophysics is very much a science of the twentieth century. Once again, this is because progress in science was intimately linked with, and built on, progress in technology — astronomers could not find out how stars work until they had the tools to do the job. Two key developments in the nineteenth century paved the way for people like Eddington to find out just what it was that made the stars tick. The first breakthrough is such a common part of everyday life today that it is hard to think of it as a scientific revolution — but it was at the time. Photography was invented at the end of the 1830s and was almost immediately applied (after a fashion) to astronomy. The brightness of the Sun made it an obvious subject for early photographers, and the first daguerreotype showing the disk of the Sun was obtained by two physicists in Paris in 1845. As photography developed during the second half of the nineteenth century, the plates used became sensitive enough (fast enough, in photographers' jargon) to record images of stars through telescopes at night. This had enormous scientific repercussions. First, even in studies as simple as the investigation of binary stars, it meant that astronomers no longer had to rely on drawings when comparing the orbits of stars from one year to the next, or one decade to the next. There was always a nagging doubt that measurements and drawings done by earlier astronomers might have been recorded incorrectly. Photography removed that particular doubt. Second, as emulsions became faster and capable of showing more detail, photographs began to reveal stars and other faint objects that could never be seen by the human eye, even with the aid of a telescope.

When you look at something, even through a telescope, the human eye soon becomes saturated and can see objects only down to a certain brightness (or faintness). If you don't see a faint star straight away, then (assuming your eyes have adapted to the dark) you will not see it at all, even if you spend hours staring through the telescope. With photographs, though, every bit of light that falls on the photographic plate, or film, adds to what went before. The longer you expose the photograph, the fainter the objects recorded. This almost literally opened up a new universe for study. But it still wasn't the most important feature of astrophotography. The greatest success of astronomical photography, the cornerstone on which the whole of astrophysics rests, came when it was combined with another scientific breakthrough of the mid-nineteenth century, the development of spectroscopy.

Spectroscopy involves the analysis of light from a star (or anything else) to provide information about what the object emitting the light (or, indeed, a gas that is absorbing the light) is made of. It can also be used, thanks to the Doppler effect, to determine how the source of the light is moving. Without those two pieces of information, there would be no science of astrophysics.

The name spectroscopy comes from *spectrum* — the familiar pattern of colored light that you see when white light is passed through a prism (or, indeed, in a rainbow itself). Like so many things in physics, the spectrum was first studied properly by Isaac Newton, who showed that white light is a mixture of the different colors that are split apart by the prism (red, orange, yellow, green, blue, indigo, and violet) and that if these colors are then recombined into a single beam of light, they appear, once again, as white light. We now explain this in terms of the wavelength of the light — red light has the longest wavelength of any of the colors of the spectrum, and violet the shortest. In a sense, spectroscopy is all about analyzing the color of light from different sources — but it does so in much more detail than simply looking at the familiar colors of the rainbow.

When the rainbow pattern produced by light sent through a prism is magnified, it turns out that there are many sharp lines in the spectrum — these lines can be dark or bright. The first person to notice this, early in the nineteenth century, was an English physicist and chemist, William Wollaston, who passed light from the Sun through a prism and saw many dark lines in the magnified spectrum. But he didn't follow through with his discovery and died in 1828, leaving its further development to others. Shortly after Wollaston's investiga-

Spectral lines. In this spectrum, the pattern of lines (the "cosmic barcode") shows up clearly. Shorter wavelengths (corresponding to blue light) are to the right, longer wavelengths (red light) to the left.

tion, the German physicist Josef von Fraunhofer also noticed the dark lines in the spectrum of the Sun, which he studied in the second decade of the nineteenth century — but he died two years before Wollaston, so the development of these ideas passed to the next generation of scientific researchers.

The key developments were made in Germany by Robert Bunsen and Gustav Kirchoff, in the 1850s and 1860s. This was the same Robert Bunsen associated with the famous piece of laboratory equipment that bears his name, even though he did not invent it; the basic burner was invented by Michael Faraday, in London, and the design was improved by Bunsen's assistant, Peter Desdega, who marketed it under Bunsen's name. But the link between Bunsen and the burner is important and relevant, because the spectra that Bunsen and Kirchoff studied were obtained by heating different substances in the clear flame of a Bunsen burner and then analyzing the light they emitted using spectroscopy.

In the spectrum of light from the Sun or a star, the dark lines show up in profusion, some thinner and fainter, others thicker and darker. Fraunhofer counted 574 lines in the spectrum of light from the Sun, each at its own precisely determined wavelength, and found many of the same lines in light from Venus (which is simply reflected sunlight, so this is not very surprising) and from many stars (which is much more interesting, since they shine by their own light).[2] The lines resemble, in quite striking fashion, the pattern of lines on a barcode today—and they are as distinctive as a barcode, because they tell you exactly what the object producing the lines is made of. The key discovery made by Bunsen, and followed up by Kirchoff, was that each element produces its own characteristic set of lines in the spectrum, as distinctive and unambiguous as a fingerprint. When the substance is hot, it produces bright lines of emission; when white light is passed through a cold gas, the result is a spectrum in which there are dark lines of absorption. But for a particular gas (say, hydrogen), the bright lines produced when the gas is hot are in exactly the same place in the spectrum (that is, at exactly the same wavelengths) as the dark lines seen when white light passes through the cold gas.

The distinctive orange light seen in street lamps, for example, is caused by a trace of sodium in the tubes of the lights that has been energized by the electric current flowing through the tube. Hot sodium always radiates energy at two well-defined wavelengths in the yellow-orange part of the spectrum, producing two bright yellow-orange lines in the spectrum. This is a particularly appropriate example, since in 1859 Kirchoff made the first identification of the presence of any element outside the Earth when he found the characteristic sodium lines (in this case, as dark lines of absorption) in the spectrum of light from the Sun.

Once the relationship between the chemical composition of a sub-

---

2. Using modern techniques, astronomers can identify more than fifteen *thousand* lines in the solar spectrum.

stance and its spectrum was appreciated, and once chemists had studied many different substances in this way in the laboratory, using the heat of Bunsen burners to take the spectroscopic fingerprints of different elements, they could immediately (as the example of the discovery of sodium in the Sun highlights) identify which elements were present in the Sun and stars by identifying which spectral lines were seen in the light from those stars — provided the stars were bright enough for their spectra to be studied in this way. They could use this tool to find out what stars were made of, even though the explanation of why different elements have unique spectra came later.

Even though the surface of the Sun has a temperature of about 6,000°C, the lines show up as a dark forest in the solar spectrum, not as bright lines, because the gas making those lines, just above the visible surface of the Sun, is cooler than the visible surface and absorbs energy as the light from the Sun passes through it.

It is the combination of spectroscopy and photography that made it possible to tell what stars are made of. Light from a star passes through a telescope and then into a prism (or onto a diffraction grating, which does the same job), where it is spread out into a tiny spectrum that is photographed with a long exposure time to bring out the details. In the early days, this process was incredibly difficult in practice, because the faint light of a single star is fainter still when spread out in this way, and photographic techniques were barely able to provide spectra for even the brightest stars. But it worked, and as the decades passed astronomers obtained spectra for fainter and fainter objects.

One interesting feature of this work was that in the nineteenth century nobody knew how the lines in the spectra were produced. They didn't have to, however, because they knew that if you looked at the spectrum of sodium, for example, you always saw the same two lines and that nothing else ever produced those particular two lines at those two precise wavelengths; so when those two lines appear in the light from the Sun or a star, you know that sodium is present in that object (or at least on its surface). The same sort of empirical argument holds for every other element.

The explanation of how the lines are produced came early in the twentieth century, when quantum theory was developed. A simple way to picture what is going on is to imagine that the electrons in an atom are in some sense in orbit around the nucleus, like the planets orbiting the Sun. If an electron jumps down from one orbit to another orbit with lower energy (very crudely speaking, as if Mars jumped into the orbit of the Earth), energy is released in the form of light—a precise amount of energy, corresponding to a precise wavelength of light, determined by the spacing between the allowed electron orbits (or energy levels). This effect, repeated in many, many atoms of the same element, produces a bright line in the spectrum, with a wavelength corresponding to the energy difference of the two orbits. Similarly, if the same amount of energy is absorbed by an atom, an electron jumps up to a higher energy level (as if jumping from the orbit of the Earth to the orbit of Mars); repeating this process in many, many atoms of the same element produces a dark line in the spectrum. Each kind of atom has a unique set of electron energy levels, so each element has a distinctive set of spectral lines. It was the explanation of the lines in the spectrum of hydrogen in quantum terms (by Niels Bohr, in the second decade of the twentieth century) that first made people take notice of quantum theory and accept that it had something useful to say about how atoms work. But still, pleasant though it is to have some idea of how the lines in spectra

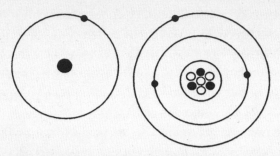

Niels Bohr explained the existence of lines in the spectrum by developing a model of the atom in which electrons orbit around the central nucleus. A sharp spectral line corresponds to the energy associated with an electron jumping from one orbit to another — if it jumps inward, energy is released to make a bright line; if it jumps outward, energy is absorbed to make a dark line. The model is only an approximation. What matters is that because each kind of atom has a unique arrangement of atoms, each element produces a unique spectral signature. The two atoms represented here are hydrogen (on the left) and lithium-7 (on the right).

are produced, it remains true that you do not need to understand quantum physics to be sure that a particular barcode pattern of lines in the spectrum of light from a star tells you what that star is made of. And then there is the Doppler shift. As I mentioned earlier, if light from a star is shifted toward the blue end of the spectrum, that is a sign it is moving toward you; if the light is shifted toward the red end, the star is moving away from you. But how do you know? Because what is actually observed across the rainbow spectrum is the pattern of spectral lines, and what is actually measured when studying the Doppler effect is how far these lines are displaced, compared with their precise positions in the spectrum (their precise wavelengths) under laboratory conditions here on Earth.

Christian Doppler actually predicted the Doppler effect, for sound waves in air, in 1842. A year later, in what must have been one of the most entertaining public scientific experiments ever carried out, he tested his predictions by having a train pull an open railway carriage full of trumpeters—all blowing away at the same note for all their might—past a group of musicians who all had perfect pitch, and so were able to record the exact change in pitch of the trumpet note as the railway carriage went by them. Doppler realized that the effect would also apply to light, but didn't go into details; the first clear account of the theory as it applied to light was set forth by the French physicist Armand Fizeau in 1848—a full ten years before Bunsen and Kirchoff got to grips with spectroscopy. So an understanding of the Doppler effect was ready and waiting when the technology of spectroscopy became good enough for the technique to be used to provide information about the velocities of stars—including information about the speed at which stars in binary systems are moving around each other, thereby helping explain stellar masses as well. Everything began to fit together in the second half of the nineteenth century.

It is because all of these techniques were coming together at the end of the nineteenth century and in the beginning of the twentieth century that astrophysics is so much a science of the twentieth century. But there was still one huge mystery, which had become more puzzling as the nineteenth century wore on—and which shows you how far astronomers still had to go, even at the beginning of the twentieth century, before they could have any hope of claiming that they understood the workings of the Sun and stars. Nobody knew how the Sun and stars could have kept shining for as long as geological evidence and the long evolutionary history of the Earth would suggest.

Although, as we have seen, advances in technology such as photography and spectroscopy were essential before astrophysics could get going, it was no less important that physicists and astronomers develop a sound theoretical understanding of how the laws of physics applied in the world at large. Newton's laws had been known for a couple of centuries or so, but the major theoretical understanding that developed in the nineteenth century concerned thermodynamics, the study of heat and of the way heat in particular and energy in general move around within a physical system and from one system to another. Some of the key insights, especially in the context of the age of the Sun, came from one of those unsung heroes of science whose work was largely ignored during his lifetime but appears outstanding in hindsight. Julius von Mayer was a newly qualified German physician who in 1840, at the age of twenty-six, took a post as ship's doctor on a vessel sailing to the East Indies. In those days, routine bleeding was still an accepted form of medical treatment (this was often carried out with enthusiasm in the tropics, where it was felt that letting a little blood from a perfectly healthy person would help counteract the debilitating effects of tropical heat). Mayer was well aware of the idea (which had been proposed by Antoine Lavoisier in the eighteenth century) that warm-blooded animals such as ourselves keep warm by a form of slow combustion within the body, in which food is combined with oxygen from the air (and is "burned"); he also knew that bright red oxygen-rich blood is delivered from the lungs to the body through arteries, while darker, purplish-red blood, depleted of oxygen, is returned to the lungs through the veins. Surgeons carrying out a bleeding were always careful to open a vein, not an artery, because the blood in the arteries is under greater pressure and bleeding from an artery is harder to stop than bleeding from a vein. But when Mayer bled a sailor in Java, he was astonished to find that his veinous blood was almost as brightly colored as normal arterial blood—indeed, at first he thought he had nicked an artery by mistake. He soon found that the same was true of the blood of all the crew, and of his own blood. Mayer quickly leaped to the correct conclusion that people had more oxygen in their veinous blood when they were in the tropics than when they were in Europe, because they were being kept warm by the heat of the Sun and so did not need as much oxygen in burning food in their muscles. He concluded that all forms of heat and energy are interchangeable — muscular exertion, the warmth of the body, heat from the Sun, and even other forms of energy such as burning coal are different forms of a single phenomenon — and, most important, that heat, or energy, is never created but is only transformed from one form into another.

When Mayer returned to Germany in 1841, he settled into general medical practice but learned a little physics on the side; he eventually developed his ideas about thermodynamics and published the first scientific papers drawing attention to the interchangeability of different forms of energy. He even published a thoughtful discussion of the heat source of the Sun, which we shall come to shortly. But his work was ignored, and when other people began to publish similar discoveries and receive the acclaim that should have been his, Mayer became so depressed that he attempted suicide (in 1850) and spent several years in a mental institution. Happily, however, his work did begin to get some recognition, he recovered his health, and he lived until 1878.

But another unsung pioneer of thermodynamics wasn't so lucky. The Scot John Waterston, a near contemporary of Mayer, was a civil engineer who worked on the burgeoning railway system of England in the 1830s, before moving to India as a teacher of cadets in the service of the East India Company. He saved hard, retired early in 1857 (at the age of forty-six), returned to England, and devoted himself to research. His main interest was thermodynamics, a subject he had been studying in his spare time; for years he had been submitting papers for publication, without success. In 1845 Waterston sent his work outlining his key insight into the way energy is distributed among the atoms or molecules of a gas from India to the Royal Society in London, but the "experts" at the society were unimpressed and not only decided not to publish the paper but lost it — the only copy in existence, since Waterston had neglected to keep a duplicate (this seems incredibly foolhardy, even in those days before photocopiers, given that the paper representing his life's work to date was being sent halfway round the world).

Like Mayer, Waterston became ill and depressed at his lack of recognition. On 18 June 1883 he walked out of his house and was never seen again. In 1891, though, Waterston's "lost" paper was found in the vaults of the Royal Society, and its importance was immediately appreciated by a new generation of physicists; it was published in 1892 — too late to do the author any good.

Waterston is relevant to our story because he had the same insight that Mayer had about the way heat is generated in the Sun. Both of them realized (ahead of their time; many other scientists were reluctant to accept the evidence at first) that, by any human timescale, the Earth must be very old indeed for the forces of nature to have done the work of raising up and wearing down mountains and so on. After Charles Darwin published his theory of evolution by natural selection in 1859, the age problem was thrown into even sharper focus, since natural selection also requires an enormous span of time to do the job of producing the variety of forms of life on Earth from some simple common ancestor. In the second half of the nineteenth century, a debate raged between the geologists and evolutionists on one side, who argued that the Earth (and therefore the Sun) must be hundreds or even thousands of millions of years old, and the physicists on the other, who argued that there was no known physical mechanism which could have kept the Sun shining for all that time.

The crowning point of the physicists' argument was produced independently by Hermann von Helmholtz in Germany and by William Thomson (later Lord Kelvin, and usually referred to by that name) in England. But they had both been preempted by Mayer and by Waterston, who had each realized that no form of chemical energy (such as burning coal) could have kept the Sun hot for more than a few thousand years and had each suggested the only alternative energy source known to nineteenth-century science — gravity.

Kelvin deserves a major share of the credit, though, because he eventually developed the idea in its most complete form. Gravity is a potential source of energy because anything falling down toward a body such as the Sun or the Earth under the influence of gravity moves faster and faster until it hits the surface — in this case, the surface of the Sun. There, its energy of motion (kinetic energy) is converted into heat energy by the impact (the same conversion of kinetic energy into heat energy explains why the brakes of a car get hot when they are used to stop or slow the car; this is dramatically obvious in the glowing disk brakes of Formula One Grand Prix cars). Both Mayer and Waterston suggested that the Sun might be kept hot for millions of years if it were "fueled" by a steady supply of asteroids (lumps of cosmic rubble a few kilometers across) falling onto it from space.

The amount of matter that would have to fall onto the Sun each year to do the trick isn't actually that extreme — only about 1 percent of the mass of the Earth per year would suffice. But, even leaving aside the question of where all that stuff could come from, if you add it up over millions (or even thousands) of years, the effect on the Sun would be quite noticeable. As the mass of the Sun increased at this steady rate, its gravitational pull on the planets would also increase, tightening its grip on the Earth and gradually shrinking the length of the year. Since we know from the records of ancient eclipses that the orbit of the Earth and the length of the year have stayed the same for thousands of years, this simple version of the gravitational heating hypothesis can be discarded. But there is an improvement on this idea that does give the Sun a potential lifetime of a few tens of millions of years, and this is the ultimate refinement of the gravitational heating scenario that was eventually developed by Helmholtz and Kelvin.

Within the context of thermodynamics, heat is understood in terms of the motion of the atoms or molecules that make up a substance. If these particles are moving faster, the object is hotter. Each particle has its own kinetic energy, and on a tiny scale it responds to gravity in the same way that a large lump of rock does. If it falls toward the center of a massive object, it gains kinetic energy from the gravitational field, and that means it moves faster. This is true even if the atoms and molecules are part of the massive object. So if the whole Sun were to shrink inward a little bit, all of the particles that make up the Sun would move a little faster, generating heat. For the Sun to keep shining at the rate we see today, radiating heat out steadily into space, it would have to shrink at a rate of only 50 meters per century—far too little to have been noticed by nineteenth-century astronomers. The orbits of the planets would not change, because the mass of the Sun, and therefore its gravitational pull, would stay the same.

There is nothing wrong with the physics—but there was still a big problem with the package of ideas, because according to this picture the Sun would shrink away entirely in about 20 million years, and by the time Kelvin had developed this argument in its ultimate form, in

1887, the geologists and evolutionists were saying that even this enormous span of time was far too little for their needs. The only way that the Sun could have generated heat for long enough to explain the evidence of the geologic record and the evolution of life on Earth was through a source of energy unknown to nineteenth-century science. That source of energy, we now know, lies in the nucleus of the atom — but the atomic nucleus would not be identified until the first decade of the twentieth century. The way ahead was pointed out, very neatly at the end of the nineteenth century, by the American geologist Thomas Chamberlin, who wrote in the journal *Science* in 1899:

> Is present knowledge relative to the behavior of matter under such extraordinary conditions as obtained in the interior of the sun sufficiently exhaustive to warrant the assertion that no unrecognized sources of heat reside there? What the internal constitution of the atoms may be is yet open to question. It is not improbable that they are complex organizations and seats of enormous energies. Certainly no careful chemist would affirm that the atoms are really elementary or that there may not be locked up in them energies of the first order of magnitude. No cautious chemist would . . . affirm or deny that the extraordinary conditions which reside at the center of the sun may not set free a portion of this energy.

Chamberlin, as we shall see, was right on target. But before this could become clear, physicists had to begin to understand those "seats of enormous energies," and astronomers still had a fair bit to do in classifying the stars and working out their family relationships to one another.

4

Inside the Stars

Scientific history, like much of history, is often told in terms of personalities. We learn who made the key discoveries and inventions, and when; the implication, though rarely stated, is that the course of scientific history might have been very different if great individuals like Isaac Newton, Charles Darwin, Marie Curie, or Niels Bohr had never lived. But this is a false impression. As I have already tried to make clear, the progress of science is inextricably linked with the progress of technology, and in addition scientific advances build on what has gone before. It is inconceivable, for example, that Isaac Newton could have come up with Albert Einstein's theory of relativity, because he had neither the knowledge about the nature of light on which Einstein built nor the mathematical techniques that were developed in the nineteenth century and that provided just the tools Einstein needed for his description of the interrelationship between space and time.

Scientific advances tend to be products of their time, and if one scientist hadn't made a particular discovery, then almost certainly another scientist would have done so at about the same time. The classic example of this is the theory of evolution by natural selection. Charles Darwin's great achievement is widely regarded as the most important scientific idea of all time — but it was discovered in exactly the same form, building on exactly the same body of earlier work, by another naturalist, Alfred Russel Wallace, soon after Darwin made his breakthrough. Darwin had kept his ideas secret, not least because he worried about their effect on his wife, a traditionally devout Christian; he published them only when Wallace sent a resume of his own, identical theory to Darwin asking for his opinion of it. If Darwin had never lived, we would probably now regard Wallace's theory of evolution by natural selection as arguably the most important scientific idea of all time.

It is very, very seldom that you can point to a major development in science and say that it depended on the existence of a unique genius. The only exception I can think of is Isaac Newton himself, who really established the whole scientific method at the end of the seventeenth century. Without Newton, quite possibly the whole development of the physical sciences might have been delayed by a generation. But my point about the logical way in which science develops from previous scientific thinking, although with the aid of new technology (which has itself been produced by a better understanding of how the world works), is made particularly clearly by the way in which astronomers discovered — or invented — the most important diagram in the whole of astrophysics, a tool on which our entire understanding of what goes on inside stars is based.

The diagram is called the Hertzprung-Russell diagram (or HR diagram), for the very good reason that it was discovered independently, at about the same time, by two astronomers (Ejnar Hertzprung and Henry Russell) working on different continents. The HR diagram plays the equivalent role as a foundation stone of astrophysics that the periodic table of the elements (itself, incidentally, discovered by several scientists independently of one another) plays as a foundation stone of chemistry. The periodic table is based on observations of the properties of the chemical elements, and tells us how the different elements are related to one another, and the theorists tell us why—a good theory, or model, of atomic structure must explain the periodic table. The HR diagram tells us how the different kinds of star are related to one another, and the theorists tell us why—a good theory, or model, of stellar structure must explain the HR diagram. And it is no surprise that two people separately came up with the idea of the HR diagram early in the twentieth century, because it was only at the end of the nineteenth century, as we have seen, that observations of things like the colors and absolute brightnesses of stars became accurate enough for astronomers to be able to classify the stars in this way.

The colors of the stars come into the story because color is related to temperature. At its simplest, this relationship is familiar from our everyday experiences. Remember those disk brakes that glow red hot? If they were even hotter, the same disks would glow blue-white; if they were a little cooler, they would radiate invisible infrared radiation but look black to our eyes. In the same way, a red star is cooler at its surface, where the light is coming from, than a white star is, and an orange-yellow star like the Sun lies somewhere in between. But astronomers can do better than that. By measuring exactly how much energy is coming from a star at a set of different wavelengths (usually three precisely chosen wavelengths, although it can be done, less accurately, with two), they can tell to a very high accuracy just how hot the surface of that star is. It is this surface temperature that is compared with the absolute brightness of the star (which depends on knowing its distance) in the HR diagram — because color is involved, and magnitude is another word for brightness, the HR diagram is sometimes called a "color-magnitude" diagram.

The Dane Ejnar Hertzprung was the first person to try to relate the colors and brightnesses of stars in a systematic way, and he published papers on this in 1905 and 1907. He found that blue and white stars are always intrinsically bright, but that some orange and red stars are bright while others are faint. In 1911, he published the first graphs relating colors and magnitudes of stars, the first examples of what we would now call HR diagrams. But all this work was published in rather obscure journals which were not widely read — certainly not by astronomers in the United States — so when Henry Russell, a senior astronomer at Princeton University, noticed the same relationship and published similar graphs in 1913, he did so without knowing anything about Hertzprung's work.

You can immediately see the power of the color-magnitude diagram by looking a little more closely at one of the first discoveries Hertzprung made, the fact that orange and red stars seem to come in two different varieties. If color depends on the surface temperature of a star, how can two stars with the same color have different brightnesses? Because some stars are big and some are small. The temperature of a star tells you how much heat is escaping across each square meter of the star's surface. So if one star has a hundred times more surface area than another, it will be a hundred times brighter, even though they both have the same surface temperature, and therefore the same color. You can even turn this around, if you know the absolute brightness and the color (temperature) of a star, to work out its size.

The most striking feature of the HR diagram, though, is that most

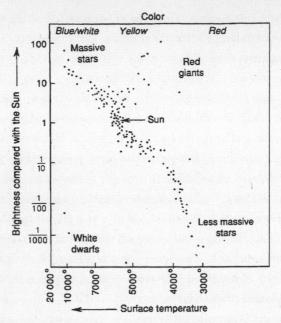

The Hertzprung-Russell diagram of stars in the neighborhood of the Sun.

stars follow the simple rule that brighter stars are indeed hotter than fainter stars. For historical reasons, although the brightness of a star is measured up the vertical axis of the graph in the usual way, temperature is plotted backward, so that it increases from right to left along the horizontal axis; this means that hotter stars are on the left and cooler stars are on the right. In this kind of plot (the one shown here is for stars in the neighborhood of the Sun, within about 70 light years from us), most stars lie on a band running from top left (hot and bright) to bottom right (cool and dim). This band is called the main sequence, and the Sun is a main-sequence star. Some exceptions to this rule are the stars which are both bright and cool, which means that they are big — much bigger than the Sun — and lie in the top right part of the HR diagram, above the main sequence. A star that is a hundred times bigger than the Sun is called a giant; one that is a thousand times bigger than the Sun is called a supergiant; and one that is a mere ten times the size of the Sun is called a subgiant (these dimensions refer to the sizes of the stars, not to their masses). From their red color and their size, the large stars are often called red giants (or red supergiants).

Down below the main sequence, in the bottom left of the HR diagram, there are stars which are both small and hot, with a great deal of heat flowing out across each square meter of surface, but with very little surface (compared with the Sun), so that even though they are white hot they are faint. They are called white dwarfs. And there are some stars which lie just below the main sequence and are called subdwarfs. But 90 percent of all stars lie on the main sequence. The crucial importance of what this is telling us about the internal structure of the stars, and the way they generate heat, didn't begin to emerge until well into the 1920s, when the pioneering astrophysicist Arthur Eddington, working at the University of Cambridge, collected all the available measurements of stellar masses and found a simple relationship between the mass of a star on the main sequence and its luminosity. The brightest stars (the ones at the top left of the main sequence) are also the most massive. Eddington found that there are stars with as little as one-fifth of the mass of our Sun, and stars with twenty-five times the mass of our Sun — and a main-sequence star that is twenty-five times as massive as the Sun is also four thousand times as bright as the Sun.

There is a certain logic to this, which was to be an immense help to astronomers trying to work out what goes on inside the stars. A more massive star has to burn its fuel (whatever that may be) more vigorously in order to hold itself up against its own weight. It ought,

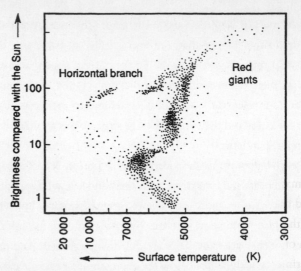

The Hertzsprung-Russell diagram of a typical globular cluster of stars. Because the largest, hottest main-sequence stars in the top left of the diagram live fast and die young, the point where the main sequence bends away to the right tells you how old the cluster is.

then, to be no surprise to find that such a star uses that fuel up more quickly than a lighter star does, and does not survive on the main sequence as long as a smaller (less massive) star does — the James Dean scenario, in which bright stars live fast and die young.

We can see exactly this process at work when we look at the HR diagrams from globular clusters of stars, groups of stars which have all been born together out of a single cloud of collapsing gas and dust, and are therefore all the same age. When astronomers look at the HR diagram for such a cluster, they find that there are no stars at the brighter end of the main sequence, and that instead there is a trail of cooler stars extending to the right in the diagram. This is known as the horizontal branch. Once we know how long it takes for stars of different masses to use up their fuel and move off the main sequence, we can work out how old a globular cluster is by measuring just where the main sequence ends and turns off toward the red giant region — but that prospect still lay far in the future in 1924, when Eddington published his mass-luminosity diagram. At the beginning of the 1920s, astronomers were only just beginning to suspect how the Sun and stars produce energy in their interiors — and they were still almost completely in the dark about what the Sun and stars are made of.

The "seats of enormous energies," deep inside the atom, that Thomas Chamberlin referred to in 1899 were already being investigated in the 1890s, although at the time nobody quite knew what it was that they were investigating. Discoveries came in a rush over a span of a couple of decades, with technology feeding experiments, experiments feeding theory, and theory encouraging the development of new experiments with more sophisticated technology. The key piece of technology involved in the breakthrough work seems unsophisticated today — an evacuated glass tube, with little or no gas left inside it, through which experimenters sent a discharge of electricity, from a metal plate (called the cathode) at one end of the tube to another metal plate (called the anode) at the other end of the tube. The whole thing is rather like the tube of a neon light, or the picture tube of a TV set. But experiments involving electricity flowing through such discharge tubes became possible only when pumps powerful enough to suck almost all the air out of the tubes became available, and it was only in the 1870s that William Crookes perfected the discharge tube.

In the 1890s, Wilhelm Röntgen was one of the many physicists studying the nature of the radiation that traveled through a discharge tube from the cathode to the anode (then called cathode rays). In 1895, he was investigating the way these rays could produce flashes of light on a fluorescent screen when he noticed another fluorescent screen, lying near his apparatus but out of the line of fire of the cathode rays, also sparking with flashes. The cause turned out to be a previously unknown kind of radiation coming from the point on the glass wall of the discharge tube where it was being struck by the cathode rays — a secondary form of radiation produced by the effect of the cathode rays themselves. He had discovered X-rays, which were soon shown to be a form of electromagnetic radiation, just like light but with much shorter wavelengths.

A couple of years later, in 1897, J. J. Thomson suggested that cathode rays are actually tiny particles, each carrying a small amount of negative electric charge, which seemed to have been chipped away from atoms (he only really proved this in 1899, but most physicists celebrated the centenary of the discovery of the electron in 1997). And both discoveries, X-rays and electrons, were very nearly made by the German physicist Philipp Lenard, twice pipped at the post as first Röntgen and then Thomson beat him to a major breakthrough. Neither breakthrough, though, could have been made by any previous generation of scientists, simply because no previous generation of scientists had good enough air pumps with which to make suitable evacuated discharge tubes.

Röntgen's discovery triggered another round of experimental work. His X-rays came from a bright spot on the glass wall of the discharge tube, where the cathode rays (electrons) made the glass fluoresce. There are several other substances which glow in a seemingly similar way, as when sunlight causes materials to fluoresce. The news of Röntgen's discovery inspired Henri Becquerel, in Paris, to investigate all the fluorescent materials he could lay his hands on, looking for anything similar to X-radiation coming from them. Becquerel took some crystals which glowed after being exposed to sunlight, and discovered that the radiation they emitted could fog a photographic plate even when the photographic plate was wrapped in two sheets of thick black paper. At first, he thought he had discovered something similar to X-rays — perhaps even that the crystals were emitting X-rays. But at the end of February 1896, Paris remained overcast for several days, while a new experiment sat waiting in a cupboard to be exposed to sunlight. This time, a dish of crystals sat on top of the double-wrapped photographic plate, with a piece of metal in the shape of a cross between the dish and the plate. When Becquerel tired of waiting for the Sun to shine, seemingly on a whim he developed the photographic plate anyway, and was astonished to find a clear image of the outline of the metal cross on it. The crystals he was working with had produced a radiation that had traveled through the black paper protecting the plate (but not through the metal of the cross) and fogged it even though the crystals had not been exposed to sunlight and were not fluorescing. Becquerel had discovered radioactivity, and soon showed that the source of this kind of radiation was uranium, one of the chemical elements present in the crystals he had been investigating.

The discovery of radioactivity was a huge puzzle, because it seemed to provide something for nothing. To make X-rays, you had to put energy, in the form of electricity, into a discharge tube to make cathode rays, and the energy of the cathode rays then triggered some effect (not then understood) in the glass of the tube, which made it glow and emit X-rays. In fluorescence, the glow of the material after being exposed to sunlight clearly came from the energy it had absorbed from the Sun — it was, if you like, stored-up sunlight. But where did the energy released in radioactivity come from?

The puzzle was presented in full force in 1903. By then, Marie and Pierre Curie, working together in Paris, had picked up from where Becquerel left off, showing that radioactivity (Marie Curie coined the term "radioactive substance" in 1898) didn't occur only in uranium, and identifying two previously unknown elements, polonium and radium, which were strongly radioactive. In 1903, the year that the Curies and Becquerel shared the Nobel Prize for physics for their work on radioactivity, Pierre Curie and his assistant, Albert Laborde, measured the amount of heat produced by radium, spontaneously, with no detectable input of energy from the world outside. Radioactivity going on inside radium makes a piece of the metal warm to the touch. Curie and Laborde found that every gram of pure radium releases enough energy every hour to heat 1.3 grams of water from 0°C to the boiling point. Radium releases enough heat to melt its own weight in ice in an hour.

This caused consternation. Some physicists seriously suggested that the law of conservation of energy, the most cherished law in science, might be being broken, with energy literally appearing out of nowhere. In 1904, Lord Kelvin, in his eightieth year, dismissed that possibility, but suggested instead that energy must be carried into the radium by some mysterious, invisible waves from outside — "I venture to suggest that somehow ethereal waves may supply energy to radium." But he was wrong, and one researcher in particular was ideally placed to pick up on the puzzle posed by Curie and Laborde and begin to probe the seats of enormous energies within the atom.

Ernest Rutherford was born in New Zealand, but at the time Becquerel discovered radioactivity, he was a research student in Cambridge, working at the Cavendish Laboratory under the supervision of J. J. Thomson (he later worked in Canada and at Manchester University, before succeeding Thomson, in 1919, as head of the Cavendish). He turned his attention to radioactivity in 1897, and soon found that the radiation discovered by Becquerel actually consists of two types of "rays," which he called alpha radiation and beta radiation, from the first two letters of the Greek alphabet. In 1900, he identified a third kind of radiation, which he called gamma radiation. Later studies showed that beta rays are actually fast-moving electrons — identical to cathode rays but carrying much more energy — while gamma rays are a form of intense electromagnetic radiation, similar to X-rays but with even higher energy. Rutherford concentrated on alpha rays (over a long period in which he also did other work), and devised a series of experiments which first showed that alpha rays are also streams of particles; then, in 1908, he showed that a single alpha particle (as it became known) had the same mass (as accurately as the experiments of the day could tell) as four hydrogen atoms, but carried two units of positive electrical charge. It was identical to a helium atom which had lost two electrons.

The modern picture of the atom as a tiny, positively charged, central nucleus surrounded by a cloud of negatively charged electrons also came from Rutherford's work with alpha particles, but not for another couple of years. This time, at Rutherford's instigation, two researchers in Manchester, Hans Geiger and Ernest Marsden, fired beams of alpha particles (produced by natural radioactive decay) at thin sheets of gold foil, and monitored the way in which the alpha particles behaved.[1] Some traveled straight through the foil without any noticeable effect, but some were deflected at a large angle, or even bounced back the way they had come, as if they had struck something solid. It was this experimental evidence that Rutherford used to devise the model of the atom as a tiny, hard central nucleus surrounded by a tenuous cloud of electrons. In modern terminology, an alpha particle is identical to a helium nucleus and consists of two protons and two

1. This is a particularly striking example of the way science progresses — barely ten years after Becquerel "discovered" radioactivity, Rutherford's team was "using" it to probe the structure of the atom.

neutrons bound together by the strong force. But the term *nucleus* was coined in this context (by Rutherford) only in 1912, soon after the experiments with alpha rays carried out by Geiger and Marsden.

Another thread of the story had already been added by work Rutherford carried out with Frederick Soddy in Canada, where he worked from 1898 to 1907. They found that radioactive decay involves the atoms of a radioactive element (we would now say, the nuclei of those atoms) being broken down to make atoms (nuclei) of a different element. When radium decays, for example, the nucleus emits an alpha particle (a helium nucleus) and is transformed into a nucleus of the gas radon. Radon is itself highly radioactive, and quickly decays further, emitting beta rays (among other things); but the details are not important here. What is much more important is the discovery, made by Rutherford, that radioactive decay always happens in accordance with a statistical law, so that for a particular radioactive element, exactly half of the atoms decay in a certain amount of time (now known as the "half-life"), and that this half-life differs for each radioactive element. Even if a half-life is much longer than a human lifetime, it can be determined by monitoring the radioactivity of a sample of a radioactive element in the laboratory for a fairly short time and measuring how the radiation starts to fade away.

The implication of these studies is that however many radioactive atoms you start with, in one half-life half of them will decay; in the next half-life half of the remainder (one quarter of the original number) will decay; in the next half-life, one eighth of the original number of radioactive atoms will decay, and so on. There is nothing magic about this—each individual atom doesn't need to "know" what the other atoms are doing. All that's required is a 50:50 chance that each atom of a particular radioactive element will (or won't) decay in one half-life. If there are enough atoms in the sample, the rest follows as logically as the fact that there is a one in six probability of getting the number 3 uppermost if you roll an honest die, no matter how often you have rolled it before or what the last number to come up was. For radium, the half-life is 1,602 years. The essential point is that this means the energy provided by radioactivity is not inexhaustible — it only seemed that way, at first, because the experiments were not sensitive enough to measure the decline in radioactivity as the original sample decays. But if you had a sample of pure radium, sealed in a radiation-proof box so none of the decay products could escape, and waited for 1,602 years, at the end of that time the heat coming from the mixture of stuff that you were left with would take two hours, not one, to melt its own weight of ice.

There was still the question of how the energy got into the radio-active nuclei in the first place; but at least physicists now knew that it was a finite reserve of energy, like a coal field or an oil well, not an inexhaustible reservoir powered by the magic of ethereal waves. As Rutherford put it as early as 1903, in his book *Radioactivity*, "the continuous emission of energy from the active bodies is derived from the internal energy inherent in the atom." That same year (the year Curie and Laborde made their measurements of the heat released by radium), Rutherford, working with Howard Barnes in Canada, was able to show that the amount of heat produced during radioactivity depends on the number of alpha particles emitted by a radioactive substance. The alpha particles collide with the atoms (actually the nuclei) of nearby material, including the other atoms of radium in the sample, giving up their kinetic energy as heat.

It was also in 1903 that astronomers began to consider the possibility that radioactivity might provide the energy source that keeps the Sun hot. The English astronomer William Wilson calculated that if there were just 3.6 grams of pure radium in each cubic meter of the Sun's volume, the energy released by radioactive decay would be enough to supply all the heat being radiated from the surface of the Sun today. The idea was taken up and promoted by the astronomer George Darwin, one of the sons of Charles Darwin, and by the end of 1903 there had developed a substantial body of opinion that the Sun's heat must come from radioactive energy. Of course, this idea was wrong; if the Sun's energy did come from the decay of radium, for example, then in 1,602 years it would be radiating only half as much energy as it does today, and in 3,204 years only a quarter as much (conversely, it would have had to be producing twice as much energy only 1,602 years ago, well within historical times). In addition, there is no spectroscopic evidence to suggest the presence of large quantities of radium (or any other radioactive element) in the Sun. But Wilson and Darwin were wrong in a sensible way (just as Kelvin and Helmholtz had been when making their calculations of solar energy production), offering the best speculation they could about the source of solar energy given the knowledge available to them at the time; and for the first time astronomers were now looking in the right place for the source of stellar energy — inside the atom. What they still lacked was an understanding of how large quantities of energy could be packed up inside the atom. But it soon came, from the work of Albert Einstein.

Einstein published his special theory of relativity in 1905. This is the theory that, among other things, says that mass and energy are interchangeable, in line with the famous equation $E = mc^2$. A mass m is equivalent to an amount of energy E found by multiplying the mass by the square of the speed of light. Since the speed of light is very large — 300 million meters per second — even a very small amount of mass is equivalent to a very large amount of energy.[2] Before the end of 1905, in a second paper on the special theory of relativity, Einstein specifically addressed the question of where the energy released in radioactivity came from, and wrote, "if a body gives off the energy L in the form of radiation, its mass diminishes by L/c^2." Although Einstein didn't do so, we can use this ratio to calculate how much mass the Sun must be losing every second in order to produce the energy pouring out into space from its surface. It is just under five million tonnes of matter every second — which seems a huge amount by human standards but is such a tiny fleabite compared with the size of the Sun that even pouring out energy at this prodigious rate for a billion years would involve the conversion of only about one-thousandth of the Sun's mass into energy. "Atomic energy" (really, nuclear energy) certainly could keep the Sun hot for long enough to explain the geological and evolutionary evidence of the long age of the Earth. But just how does nature convert that relatively small amount of m into all that E?

Astronomers began to realize that they had been barking up the wrong tree by thinking about stellar energy production in terms of radioactive decay only in 1919 — and they realized it then thanks to another experimental breakthrough, which provided crucial new evidence about the nature of atomic nuclei. A decade earlier, Rutherford had measured the masses of alpha particles and found that it is roughly the same as the mass of four hydrogen atoms. But in 1919,

2. Since some people still find it hard to accept the "non-commonsense" nature of Einstein's theory, it is worth emphasizing that this is not some wacky idea from a mad professor, but that every prediction of the theory, including the relationship between mass and energy, has been tested by experiments numerous times since 1905, and the theory has proved to be a good description of the way the world works, to many decimal places. You may not like it, but if you cannot accept it, you are in the same position as someone who believes that the Earth is flat.

Francis Aston, working at the Cavendish Lab, developed a more accurate way to measure these masses (it involved studying the way the charged particles are deflected by magnetic fields) and discovered that the mass of an alpha particle is not quite the same as the mass of four hydrogen nuclei (four protons) put together. That "not quite" explains how all stars on the main sequence of the HR diagram maintain their output of energy.

Aston's measurements showed that the mass of a nucleus of helium is 0.8 percent less than the mass of four single nuclei of hydrogen (four single protons) added together. This was still before the discovery of the neutron, and physicists weren't quite sure what nuclei were made of; but a reasonable guess seemed to be that a helium nucleus contained four protons plus two electrons to cancel out two units of the positive charge of the protons. The mass of an electron is only about half of one-thousandth of the mass of a proton, so it doesn't come into the calculation at this level, and doesn't affect the argument. Because the atomic weights of the elements are very nearly integer multiples of the atomic weight of hydrogen, it was clear that the atoms of other elements must be built up (somehow) using hydrogen as the basic building block; but the accuracy of Aston's measurements (involving other elements, as well as helium) showed that a tiny bit of mass got lost along the way. Eddington seized on the idea, and the following year, at the 1920 meeting of the British Association for the Advancement of Science (the BA), he spelled out the implications to an intrigued, if somewhat startled, audience:

> A star is drawing on some vast reservoir of energy by means unknown to us. This reservoir can scarcely be other than the sub-atomic energy which, it is known, exists abundantly in all matter; we sometimes dream that man will one day learn to re-

lease it and use it for his service. The store is well-nigh inexhaustible, if only it could be tapped. There is sufficient in the Sun to maintain its output of heat for 15 billion years. . . .

Aston has further shown conclusively that the mass of the helium atom is even less than the masses of the four hydrogen atoms which enter into it — and in this, at any rate, the chemists agree with him. There is a loss of mass in the synthesis amounting to 1 part in 120, the atomic weight of hydrogen being 1.008 and that of helium just 4. I will not dwell on his beautiful proof of this, as you will no doubt be able to hear it from himself. Now mass cannot be annihilated, and the deficit can only represent the mass of the electrical energy set free in the transmutation. We can therefore at once calculate the quantity of energy liberated when helium is made out of hydrogen. If 5 percent of a star's mass consists initially of hydrogen atoms, which are gradually being combined to form more complex elements, the total heat liberated will more than suffice for our demands, and we need look no further for the source of a star's energy.

Eddington, who virtually invented the scientific discipline of astrophysics, exactly hit the mark with the main thrust of these comments. But progress was hamstrung for years by two problems, one theoretical and one observational. On the theory side, nobody knew how two protons (let alone four) could get close enough to join together. Even granted that there must be some mysterious force which held atomic nuclei together even though the positive charge on all the protons in the nucleus was trying to blow it apart (and this was long before the strong nuclear force was even remotely understood), it was clear that this binding effect didn't extend beyond the nucleus, or it would suck everything else into one huge lump of matter, like a single giant nucleus. If two protons collided head on and touched, they might, perhaps, stick together. But since they each have a positive charge, they are repelled from each other with a force proportional to one divided by the square of the distance between them, which becomes bigger and bigger the closer and closer they get. So how could hydrogen atoms "gradually be combined to form more complex elements"? The other problem was really a misunderstanding which sent Eddington and his colleagues off on the wrong track when they tried to work out details of the kind of subatomic processes going on inside stars. In that talk to the BA in 1920, he referred to the possibility that 5 percent of a star's mass might be made of hydrogen. It is this guess, coupled with the possibility that all that hydrogen might be converted into helium, which gave Eddington the estimated lifetime for the Sun of 15 billion years. But why pick 5 percent? Because at the beginning of the 1920s astronomers thought that the composition of the Sun and stars was, broadly speaking, similar to the composition of the Earth. This was partly a kind of unconscious parochialism, the unstated assumption that other objects in the Universe are made of the same sort of stuff that we are; but it was also partly a misunderstanding of that forest of lines seen in the spectrum of the Sun, indicating a huge variety of elements present in the atmosphere of our nearest star. If anything, by the accepted standards of 1920 Eddington was erring on the high side in suggesting that as much as 5 percent of the Sun's mass might be made of hydrogen.

Both these problems were resolved in the second half of the 1920s, and that is when the study of the internal structure of the stars really took off. But in the meantime, Eddington had pointed the way by making the first calculations of the temperatures that must exist at the hearts of stars, using very simple (literally, school-room level) physics and a great deal of insight, plus the growing amount of information about the relationship between the masses and luminosities of main-sequence stars.

Eddington realized that you didn't need to know where a star got its energy from in order to get a rough idea of what was going on inside it. He also realized that the basic physical laws which describe what is going on inside a star are the laws which describe the behavior of a hot gas — one of the simplest, and best-studied, kinds of system investigated by physicists. This seems surprising at first sight, since the average density of the Sun is one and a half times the density of water, and the density at its center is many times the density of lead. But the "gas" the stars are made of isn't like the air that you breathe.

An ordinary gas is described by very simple laws and equations because it behaves like a lot of little hard spheres (the atoms) bouncing around and colliding with one another and with the walls of any container the gas is confined in. In a solid — like lead — the atoms are packed tightly together and don't do much moving around. But, as I described in Chapter 2, the nucleus of an atom is far smaller than the atom itself. When any material (hydrogen, or lead, or anything else) is hot enough, the energy of the collisions between the particles, and the effect of the electromagnetic radiation they produce interacting with the charged particles, knocks the electrons away from the atoms, leaving bare nuclei behind. The resulting hot mixture of positively charged nuclei and negatively charged electrons is called a plasma, and it behaves like a gas because now we have the *nuclei* acting like little hard spheres that bounce around and collide with one another. The difference in size between an atom and a nucleus is so great that a plasma carries on behaving like a perfect gas even at densities far greater than those at the heart of the Sun. And the laws that describe the behavior of a perfect gas tell us how hot a star with a certain mass and a certain luminosity must be inside in order to hold itself up against the inward tug of gravity.

In fact, there is a rather delicate balancing act going on. As well as the ordinary pressure caused by the particles inside a star bouncing around and colliding with one another, the particles, because they are charged, also radiate a lot of electromagnetic energy — things like X-rays and gamma rays. This radiation interacts with other charged particles in the plasma and provides an additional pressure, called radiation pressure. If a ball of gas in space collapses and gets hot inside (initially as a result of the release of gravitational energy, just as Kelvin and Helmholtz described), there are three possible fates for it. A small gas globe will not get very hot inside, the heat energy will radiate away, and it will cool down on the sort of timescale that Kelvin and Helmholtz discussed. It will end up as a cool ball of gas, rather like the planet Jupiter, or as a so-called brown dwarf, perhaps seventy times more massive than Jupiter (but only 7 percent of the mass of our Sun), a nearly star that just failed to trigger the nuclear fusion processes that keep main-sequence stars shining. At the other extreme, a large gas globe will generate so much heat as it collapses that a dense, hot plasma is created in its heart, and the combination of gas pressure and radiation pressure blows it apart, so it never settles down as a main-sequence star. But in between these two extremes is a small range of masses where the gas globe gets hot enough for the plasma to form (and, we now know, for nuclear reactions to start generating heat in its heart), but not so hot that it blows itself to bits. Only stars with masses roughly in the range from one-tenth of the mass of our Sun to a hundred times the mass of the Sun are stable in this way — and this follows, from the simple laws of gas (plasma) physics, whatever the process by which stars actually generate heat in their interiors. Gratifyingly for astronomers, when we look at the stars we do indeed find that there are none with masses less than a tenth of a solar mass, and none with masses much more than a hundred times the mass of the Sun. The Universe really does operate according to the same laws of physics that we study in laboratories here on Earth.

When you do the calculations, as Eddington did, you can even work out the temperature that must exist at the heart of any star today, if you know its mass, its luminosity, and what it is made of. The composition comes into the story because it affects the number of little hard spheres bouncing around inside the star and contributing to the pressure that holds it up. If there are fewer particles, they each have to move faster to provide the same overall pressure — which means that they have to be hotter. What matters is the number of atomic nuclei, which each count as one particle in this calculation. Since each helium nucleus is essentially made up from four hydrogen nuclei (as I explain in the next chapter), a star that is, for example, made up entirely of hydrogen would have four times as many particles bouncing around inside it as a star with exactly the same mass but made entirely of helium (and if a star existed that were made entirely of radium, which has an atomic mass of 226, it would contain only 1/226 as many nuclei as a star of the same mass made entirely of hydrogen). Other things being equal, the helium star would have to be a lot hotter in its heart than the hydrogen star (and the radium star correspondingly hotter still) in order to hold itself up against the inward tug of gravity.

Because he did not realize that stars on the main sequence are actually made almost entirely of hydrogen and helium, when Eddington did the calculation he worked out a number for the central temperature of a main-sequence star that was rather too high — about 40 million Kelvin (which for all practical purposes is the same as 40 million degrees Celsius). But this wasn't important. What was important was his discovery, from the mass-luminosity relation and the laws of gas physics, that all main-sequence stars have essentially the same central temperature. He had clearly discovered a fundamentally important feature of the internal workings of the stars. In his book *The Internal Constitution of the Stars,* published in 1926, he referred to two specific examples of stars that he had investigated, and wrote that "taken at face value [this] suggests that whether a supply of 680 ergs per gram is needed (V Puppis) or whether a supply of 0.08 ergs per gram (Krueger 60) the star has to rise to 40,000,000° to get it. At this point it taps an unlimited supply." Later in the book, he elaborated on this theme: "[A star will] contract until its central temperature reaches 40 million degrees when the main supply of energy is suddenly released. . . . A star on the main series [main sequence] must keep just enough of its material above the critical temperature to furnish the supply required." The key point, regardless of the exact temperature that came out of the calculation, is the implication that all main-sequence stars, the Sun included, get their energy in exactly the same way. Tantalizingly, the book was published just as the quantum physicists were coming up with new ideas about how particles like protons behave, which would soon explain how the fusion process overcomes the electrical repulsion between protons. Always up to the minute, in a preface dated July 1926 Eddington mentioned that "as we go to press a 'new quantum theory' is arising which may have important reactions on the stellar problem when it is more fully developed." He was right.

The key feature of the quantum theory that was developed in the second half of the 1920s, and has been a cornerstone of physics ever since, is that at the subatomic level of things like protons and electrons the quantum entities do not behave exactly like little hard spheres. They behave like a mixture of a wave and a particle (a phenomenon known as wave-particle duality). It works both ways — light, which nineteenth-century physicists would have described solely in terms of electromagnetic waves (the way I have described it so far) also behaves like a stream of tiny particles, called photons; electrons, described by J. J. Thomson in terms of little particles, also behave like waves. This is not the place to go into all the details (which I have covered in my book *In Search of Schrödinger's Cat*); but, as with the theory of relativity, very many experiments have confirmed, to very many decimal places, that the quantum world (the world of atoms and smaller things) really does behave like this. The effect is bigger for less massive particles and doesn't show up at all on the scale of things that we can see with our own eyes, like sugar lumps or rhinoceroses. What matters here is that the insight provided by quantum theory tells us that it is inappropriate to regard a proton as a tiny sphere with a well-defined edge. Instead, it is more appropriate to think of it as a concentration of mass energy and electric charge associated with a small group of waves, called a wave packet.

In 1928, a young Russian physicist who was visiting the University of Göttingen realized that this waviness of fundamental entities could explain how radioactivity worked — how alpha particles escape from an atomic nucleus during the process of radioactive decay. The puzzle is that even in a radioactive nucleus the strong force that holds the nucleus together is, according to calculations which do not take account of these quantum effects, just (but only just) too strong to let the alpha particles escape. An alpha particle which is just outside the nucleus will not feel the strong force and will be repelled from the nucleus, because both the nucleus and the alpha particle have positive charge. But inside the nucleus all the particles are held in place by the strong force, which overwhelms the electrical repulsion. It is as if the particles were lying in the mouth of a volcano — the alpha

Quantum physics reveals that entities which used to be thought of as tiny point-like particles (such as electrons and protons) should be regarded as spread out over a volume of space, rather like a short pulse of waves (a wave packet).

particles don't have enough energy to climb out of the volcano and roll away down the sloping mountainside outside.

But George Gamow realized that the waviness of an alpha particle meant that it was, in a sense, too big to fit neatly inside the mouth of the volcano. Some of the waviness would extend through the side wall of the "mountain," so that the alpha "particle" could gradually (on a timescale related to the half-life) leak through to the other side. Once there, it would "roll away" as it was repelled by the positive charge on the nucleus. This is called, for obvious reasons, the "tunnel effect." And although I have only sketched the outlines of the idea here, once again precise calculations using quantum theory do indeed predict exactly the amount of alpha radiation (the right half-lives, and so on) associated with tunneling out of atomic nuclei, such as those of radium, that we do actually observe.

This was exciting stuff for the particle physicists. But Gamow also realized that the process could work the other way around. If two positively charged protons approach one another closely enough, even though the positively charged cores of the wave packets are not in contact, the extended edges to the wave packets can overlap. This overlapping can pull the two waves together, even though simple particles approaching one another with the same amount of energy (the same speed) could never touch and come under the influence

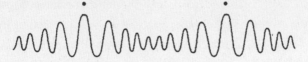

If two protons (or other nuclei) approach one another at a certain speed, the electrical repulsion caused by their positive charge will stop them touching and interacting—*if* they are particles. But if they are wave packets, the edges of the wave packets can interact over a longer range. This is what makes nuclear fusion possible at the temperatures that exist inside the Sun and other stars. The existence of stars like the Sun is confirmation of the accuracy of the quantum description of the wavelike nature of particles.

of the strong force. Its like two people swimming in the sea who reach out and hold hands, pulling themselves toward each other even though the waves are trying to move them apart. Gamow immediately pointed this out to his astronomer friends, and two of them took up the suggestion of using the tunnel effect to try to explain how nuclear fusion produces energy inside stars. But they were still handicapped by thinking initially in terms of hydrogen nuclei (protons) approaching and interacting with larger nuclei (very much like alpha decay in reverse), instead of thinking in terms of protons interacting directly with one another. The idea that hydrogen might be the main component of stars took a very long time to sink in, even though clear evidence of this also emerged in 1928, the same year that Gamow came up with the idea of the tunnel effect.

This is another example of a scientific idea emerging from new technology when the time was ripe. The first suggestion that the atmospheres of the Sun and stars were rich in hydrogen came from the work of Cecilia Payne (later Cecilia Payne-Gaposchkin), an English-born astronomer who in 1925 received her Ph.D. from Radcliffe College for her investigation of the relationship between stellar temperatures and spectra. As part of her doctoral work, she noticed, using spectroscopy, that the composition of stellar atmospheres is dominated by hydrogen. This can be seen with hindsight as one of the first indications of the overwhelming preponderance of hydrogen in the visible Universe. But Payne's Ph.D. examiner insisted that, when she published this discovery, she include a comment to the effect that the intensity of the hydrogen lines in the spectra she had studied must be caused by some peculiar behavior of hydrogen under stellar conditions rather than by a very high abundance.[3] In 1928, though, the German astronomer Albrecht Unsöld carried out a detailed spectroscopic investigation of the light from the Sun and, taking the data at face value, interpreted the strength of the hydrogen lines as implying that there are roughly a million times as many hydrogen atoms there as there are atoms of anything else. Just a year later, a young British astronomer, William McCrea, reached a similar conclusion, using a different spectroscopic technique.

The triple discovery was soon welcomed by astrophysicists, because it showed that there was plenty of hydrogen in the Sun, certainly enough to provide the energy required to keep it shining in much the same way for billions of years, through, as Eddington had suggested, the conversion of hydrogen nuclei into helium nuclei. For two decades, though, nobody realized that these studies of the atmosphere of the Sun were telling astronomers that the "interior" of the

3. In the scientific paper based on her doctoral dissertation, she actually said, using the words put in her mouth by her examiner in reference to hydrogen and helium, "the enormous abundance derived for these elements in the stellar atmosphere is almost certainly not real." Payne was a first-class astronomer, even as a student, and felt that she had found out something important about what the stars are made of; the fact that her examiner (Henry Norris Russell) had trouble believing what the spectra were telling her shows how difficult it was for astronomers to come around to the idea that stars are not made of essentially the same stuff that the Earth is made of.

Sun was dominated by hydrogen (with the implication that other stars are also made chiefly of hydrogen, rather than just having a lot of hydrogen in their atmospheres). But in spite of this confusion, and several false steps along the way, by the end of the 1930s, little more than ten years after the discovery of the tunnel effect, the astrophysicists had found not one, but two ways to convert protons, four at a time, into helium nuclei in the hearts of main-sequence stars.

5

Cycles and Chains in Stars

George Gamow published his discovery of the tunnel effect in 1928. The following year, two young physicists, Robert Atkinson and Franz Houtermans, published the first calculations of how the tunnel effect might operate inside stars. Their paper began with the words "Recently Gamow demonstrated that positively charged particles can penetrate the atomic nucleus even if traditional belief holds their energy to be inadequate," and went on to calculate the kind of nuclear reactions that might be involved. That opening sentence both sums up the extent of the leap that Gamow's work inspired and also demonstrates how far astrophysicists still had to go in unraveling the secrets of nuclear fusion inside stars. For although Unsöld's work on the composition of the Sun's atmosphere had been published in 1928, and McCrea's contribution appeared in the same year as the paper by Atkinson and Houtermans, they were, as that opening sentence makes clear, still thinking in terms of something like alpha decay in reverse, with simple particles penetrating into the nuclei of massive elements. Following Arthur Eddington's lead, they suggested that the fundamental process which provided the energy released by stars was indeed the conversion of four protons into one helium nucleus (one alpha particle). But they didn't suggest that this happened directly; rather, they used the analogy of a cooking pot. The "pot" would be a heavy nucleus at the heart of a star like the Sun, which absorbed the ingredients (four protons and two electrons) from its surroundings one at a time, cooked them up to make a helium nucleus, and then spat it out, through alpha decay, ready to repeat the whole process.

The most important feature of the work by Atkinson and Houtermans was that they put numbers into the calculations, numbers based on the growing understanding of the tunnel effect that was emerging from a combination of experimental studies of processes like alpha decay and the new quantum theory of the late 1920s. Since stars are the biggest things that we can see with our own eyes, and quantum physics deals with entities much smaller than atoms, the fact that quantum physics explains how stars work is an impressive demonstration of the way our scientific understanding of the Universe is linked on all scales — and impressive confirmation that the whole scientific endeavor is on the right track.

If you think of the electrical repulsion between two positively charged particles that are approaching each other as a physical barrier like a hill, it is fairly obvious that the hill will be higher and harder to penetrate if the particles have more positive charge, and also that it will be easier for a particle to get through the barrier if it is moving faster. At a particular temperature, lighter particles move faster than heavier particles. From the calculations that people like Eddington had already made about the structure of stars, Atkinson and Houtermans knew roughly what kind of temperatures were involved, and what kind of densities and pressures existed at the hearts of stars. So they knew how fast the particles there were moving, and how hard the particles would be colliding with one another. They showed that even with allowance made for the tunnel effect, under the conditions which existed inside main-sequence stars only the fastest-moving particles with the smallest positive charge (in other words, protons, nuclei of hydrogen) could penetrate the barriers. The processes that keep stars shining (at least, main-sequence stars) had to involve hydrogen directly and could not solely involve collisions between pairs of massive nuclei, which then rearranged themselves and spat out alpha particles.

It's quite startling to appreciate just how difficult it is for a main-sequence star to obtain energy in this way. Every time I look at the numbers myself, I am amazed at how feeble a process the mechanism of stellar energy generation is. It sounds bizarre — after all, the Sun is the most profligate source of energy in our immediate neighborhood. How can it possibly be described as feeble? But consider this. First, some of the protons inside a star like the Sun are moving faster than others — the overall distribution of speeds depends on the temperature, which tells us both what the average speed of the particles is and (very precisely) what proportion are moving a set amount (say, 20 percent) faster or slower than the average speed. Updating the calculation made by Atkinson and Houtermans to take account of the best modern understanding of how stars work, it turns out that (at the temperatures which exist at the heart of the Sun) two protons will fuse with each other, even with the aid of the tunnel effect, only if one of them is moving at least five times faster than the average speed. Even for such fast-moving particles, fusion occurs only if the collision is almost perfectly head-on. If there is any angle at all between the tracks of the two particles, they strike each other a glancing blow and proceed on their way. For the Sun itself, just one proton in every hundred million is moving fast enough to be able to penetrate the barrier around another proton, and just one collision out of every ten billion trillion (1 in 10^{22}) results in fusion. This means that on average an individual proton will spend 14 billion years bouncing around inside the Sun, colliding with and recoiling from other particles, before a rare head-on collision enables it to fuse with a partner. Even at the heart of the Sun, nuclear fusion is an extremely rare process as far as individual protons are concerned. But there are so many particles inside the Sun that enough protons to provide 616 million tonnes of hydrogen are involved in such collisions each second, producing enough alpha particles to make 611 million tonnes of helium, with 5 million tonnes of mass being converted into energy. This is such a tiny proportion of the Sun's mass, though, that after 4.5 billion years as a main-sequence star, only 4 percent of its initial stock of hydrogen has been converted into helium.

Atkinson pushed forward the development of the theory of fusion inside stars on his own in the early 1930s, as Houtermans developed other interests. He looked at different kinds of nuclear interactions, with hydrogen nuclei penetrating nuclei of other elements, using a mixture of theoretical calculation and experimental data. But even though he had shown, by 1936, that under the conditions prevailing inside the Sun, the most common nuclear interaction is one in which two protons come together to make a nucleus of deuterium (heavy hydrogen), astronomers still hadn't realized that hydrogen forms by far the bulk of the Sun.

The reason was due to an unfortunate coincidence. After Unsöld and McCrea had shown that there must be a lot of hydrogen in the Sun (and, by implication, in other main-sequence stars), astrophysicists reworked the pioneering calculations of stellar structure that had been made by Eddington and were refined in the 1930s by the Indian-born astrophysicist Subrahmanyan Chandrasekhar. The key to this next step in understanding stellar structure was the number of electrons inside a star — or, strictly speaking, the number of electrons per nucleon, where the term *nucleon* refers to either protons or neutrons. This is important because of the way electromagnetic radiation interacts with charged particles. Part of the pressure that holds a star up comes from this interaction, and the more electrons and protons there are around, the more effective this radiation pressure will be. If a star were entirely made of hydrogen, there would be one electron for every proton — one electron per nucleon, since there would be no neutrons at all. If it were entirely made of helium (alpha particles each containing two protons and two neutrons), there would still be one electron for every proton, but now only half an electron per nucleon, since the neutrons would have no electronic partners. The number of electrons per nucleon goes down as the proportion of heavier elements increases, and this affects the stability of the star through the pressure associated with electromagnetic radiation (as well, of course, as the effect on the structure of the star of having the neutrons and protons crammed together in heavy nuclei, which I mentioned earlier). And if you think astrophysicists were a bit slow to latch onto all this after Eddington had pointed the way, remember that the neutron was not discovered until 1932, so they actually moved on from that discovery with breakneck speed.

Once people knew that there was a lot of hydrogen in the Sun, it was natural to try to work out what proportion of hydrogen to heavier elements would be needed to make a star like the Sun stable. Unfortunately, because of a tradeoff between the different factors affecting the stability of a star, this question has two answers. It turns out that a star with the mass and luminosity of the Sun (indeed, any similar main-sequence star) will be stable if at least 95 percent of its mass is made up of hydrogen + helium put together. But such a star will also be stable if it is made up of 35 percent hydrogen and 65 percent heavy elements.

Before about 1928, it had generally been assumed that the Sun, like the Earth, was largely made of heavy elements. So in the 1930s, when the calculations offered astrophysicists a choice of stellar models, one with 65 percent heavy elements and one with less than 5 percent heavy elements, it isn't too surprising that they unanimously plumped for the model with 65 percent heavy elements and dismissed the alternative model as an unimportant fluke of the way the calculation worked out. The mistake wouldn't really begin to be rectified until the end of the next decade, and would not be properly corrected until the 1950s, even though the astrophysicists had at last put their finger on the key nuclear reactions that go on inside the stars by the end of the 1930s.

Once again, George Gamow acted as a catalyst for the new developments. In April 1938 he organized a conference in Washington, D.C., so that astronomers and physicists might come together to discuss the problem of energy generation inside stars. The puzzle was to find a set of nuclear reactions, coupled with a more detailed composition for the stellar model, that would produce energy at just the right rate to keep a star like the Sun shining steadily at its present output for billions of years. Atkinson (and others) had been trying for years to find the right set of reactions, but everything they came up with was either too fast or too slow. For example, if there were a lot of lithium inside the Sun, then hydrogen nuclei would combine with lithium nuclei eagerly, even at a temperature of only 15 million degrees, building up nuclei of unstable beryllium, which would each rapidly split into two helium nuclei. The series of reactions converting hydrogen into helium would occur so quickly, and release so much energy in such a short time, that it would blow the star apart. On the other hand, if most of the star were made of oxygen nuclei, although protons could interact with these nuclei to release energy, they would not release enough energy to keep the star shining at the brightness the Sun does today. So the star would contract, releasing gravitational energy and growing hotter inside, until it was hot enough for this process (or something else) to produce enough energy to stabilize it. Nobody at the conference could come up with a set of nuclear reactions that would be, like Baby Bear's porridge, "just right." But they set off home afterward with the puzzle buzzing around their minds — and one of them, Hans Bethe, from Cornell University, solved it.

There's a delightful anecdote about how Bethe solved the puzzle, which unfortunately bears no relation to the truth. Gamow was a great joker who loved telling stories, and if the truth was a little boring he would be happy to embellish it. So he used to tell the tale of how Bethe caught the train from Washington after the meeting, decided that it shouldn't be too difficult to solve the puzzle, and set himself the task of doing just that before the steward called the passengers to dinner. According to the legend invented by Gamow, Bethe, who always enjoyed a good meal, promised himself that he would solve the puzzle before he allowed himself to eat, and after a furious bout of scribbling calculations in pencil he found the answer exactly at the moment the steward came by with the call to dinner. But as Gamow himself admitted in his book *The Birth and Death of the Sun,* we should not give "too great credence" to the story of "the relationship between Dr Hans Bethe's famous appetite and his rapid solution of the problem of solar reaction."

In reality, Bethe did not find it quite so easy to solve the puzzle, and although he did start work on it on the train, he finished the solution back at Cornell (without missing any meals). Unknown to him, the same solution to the puzzle had already been found by Carl von Weizsäcker, in Germany, earlier that same year; but Bethe usually gets pride of place in the story, not just because von Weizsäcker had no Gamow around to publicize his achievement, but also because of something Bethe did in the summer of 1938, which I shall tell you about shortly. The process they both discovered is very much in the spirit of the pioneering work of Atkinson, because it does indeed involve nuclei of hydrogen (protons) penetrating the nuclei of heavier elements — specifically, carbon, nitrogen, and oxygen — in a multi-step process which ends with an alpha particle being ejected from the nucleus. This is exactly the kind of nuclear cooking-pot effect that Atkinson and Houtermans had speculated about back in 1929, but now with accurate numbers put in to specify the rates of the various steps in the reaction. It turned out, though, that this is not the main process which keeps the Sun itself shining, because it operates most effectively at a slightly higher temperature than the temperature at the heart of the Sun (above about 20 million degrees Celsius); such temperatures exist in the hearts of stars with at least one and a half times as much mass as the Sun, so this nuclear cycle discovered by Bethe and by von Weizsäcker is the process which keeps stars farther up the main sequence shining. But it is extremely important for our own existence, and it came first historically, so this is the right place to spell out how it works.

Because the cycle starts with carbon, it is often called the carbon cycle. Because nuclei of nitrogen and oxygen are also involved, it is also referred to either as the CN cycle or the CNO cycle. The participation of hydrogen is taken for granted, but no astronomer that I know ever refers to the cycle by the appropriate full acronym, though it really is a CHON cycle. The nuclear processes that keep many of the stars on the main sequence shining depend on the existence of exactly the same elements that are important for life as we know it, emphasizing the close relationship between life and the Universe.

It works like this. First, there has to be at least a smattering of heavier elements around inside the star. This was not seen as a problem in the 1930s, of course, since it was thought that 65 percent of a star was heavy elements; today, this tells us that we are dealing with second (or later) generation stars, stars that have been made out of material that has been at least partially processed inside other stars already. The cycle is literally that — it goes round in a loop — so you could start with any step in the process, but it is natural to start with carbon. First, a proton tunnels into a nucleus of carbon-12, which contains six protons and six neutrons already. This transforms the nucleus into unstable nitrogen-13, which is radioactive, and emits a positron and a neutrino, converting itself into the stable nucleus carbon-13.[1] If a second proton tunnels into the nucleus of carbon-13, it makes another stable nucleus, nitrogen-14. But if a third proton tunnels into the nitrogen-14 nucleus, it creates another unstable nucleus, oxygen-15, which also decays by emitting a positron and a neutrino, thus converting itself into stable nitrogen-15. Now, though, something more dramatic can occur. If a fourth proton tunnels into the nucleus of nitrogen-15, it is disrupted and promptly ejects an alpha particle, leaving behind a stable nucleus of carbon-12, just like the one we started with. The net effect of the cycle (wherever you start going round it) is that four protons have been cooked up in the nuclear pot to make one helium nucleus (an alpha particle), together with a couple of positrons and two neutrinos; along the way, at various steps in the cycle the interactions have also produced electromagnetic radiation. But we

---

1. A positron is a positively charged counterpart to an electron; when a positively charged proton ejects a positron (and a neutrino), it is left with no charge at all and becomes a neutron. Emitting a positron is exactly equivalent to absorbing an electron.

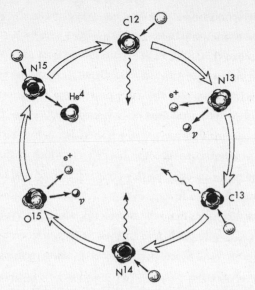

The CNO cycle, which is the main source of energy in stars slightly more massive than the Sun.

haven't "used up" any of the other nuclei involved in the cycle — the carbon and nitrogen and oxygen are still there and can be used again and again in many such cycles, while the unstable oxygen-15 is made every time it is needed, making this a plentiful source of energy even if relatively few nuclei of heavy elements are involved.

The power of this description of a stellar energy source is that each step in the cycle can be studied here on Earth — we know, from experiments, what happens when protons interact with each of the nuclei involved. So we know how fast the reactions proceed under laboratory conditions, and we can use quantum physics to extrapolate that speed to calculate how fast the cycles proceed under the conditions that operate inside stars. These are not mere hand-waving arguments, but proper quantitative calculations. Indeed, the calculations are so precise that it is possible to include secondary effects which affect the cycle. The most important of these is a kind of side loop, starting out from the nitrogen-15 link in the chain. Sometimes (a proportion of the time which can be precisely calculated), instead of emitting an alpha particle as soon as it absorbs a proton, the nitrogen-15 nucleus is transformed into a nucleus of oxygen-16, which itself absorbs a proton to become fluorine-17, which then decays by ejecting a positron and a neutrino to become oxygen-17, which itself absorbs a proton and emits an alpha particle, making a nucleus of nitrogen-14 which connects this side loop back to the main cycle. And there are other, rarer diversions from the main loop, which I won't describe in detail here. The point is that in this whole network of loops, whichever path is followed in any particular series of interactions, the overall effect is always to convert four protons into one helium nucleus, a couple of positrons and neutrinos, and energy. The rest is unchanged.

There is, though, another subtlety about this network of interactions that is especially interesting to life forms like ourselves. Everything remains unchanged *provided the process has reached equilibrium*. Each step in the network of interactions proceeds at a different rate, and this affects the nature of the overall equilibrium that is achieved. Where the reactions proceed quickly, the nuclei involved do not stay around in large quantities. But where the reactions proceed more slowly, there is something of a log jam, and the particular nuclei involved build up until an equilibrium is reached, with the same number of new nuclei being manufactured as the number of old nuclei that are being destroyed.

Think of three buckets, one above the other, each with a hole in the bottom and all being fed with a steady flow of water from a tap. The rate at which water flows out of the hole in the bottom of a bucket depends both on how big the hole is and on how much water there is in the bucket—more water makes more pressure and makes the water squirt through the hole faster. You could adjust the size of the hole and the flow of water from the tap to keep the first bucket always, say, one quarter full. That would produce a steady flow of water into the second bucket, where, by making the hole a little smaller you could ensure that the bucket was always three quarters full. Finally, the water flows into the bottom bucket, which has a larger hole in the bottom, and so is always just one-eighth full. Overall, there is a steady flow of water through the system, but as far as each bucket is concerned, there is an equilibrium.

If you turn the tap off and let the buckets empty, when you turn it back on (to the same strength, of course), the water levels in the buckets will rise back to their respective equilibrium levels, and stay there. Even if you suddenly dumped an extra liter or two of water into one of the buckets (any of them—or all of them) or scooped some out, they would soon settle back to the same equilibrium levels.

The carbon cycle operates in much the same way, and is in equilibrium, with no overall change in the number of each kind of nuclei present, when there is 5.5 percent carbon-12, 0.9 percent carbon-13, 93.6 percent nitrogen-14, and 0.004 percent nitrogen-15. This equilibrium will be achieved whatever the original mixture of these elements was when the carbon cycle started to do its work—even if, to take an extreme case, there was no nitrogen present at all to start with, equivalent to having an empty bucket. The large proportion of nitrogen 14 arises because the reaction that converts nitrogen-14 into nitrogen-15 (this absorption of a proton is called "hydrogen burning") is very slow compared with the other steps in the cycle (the nitrogen-14 bucket has only a small hole in it). So one effect of the carbon cycle, over the lifetime of a star at the more massive end of the main sequence, is the conversion of most of the carbon and oxygen which were present in the star to start with into nitrogen.[2] I'll explain where the carbon and oxygen came from in Chapter 7; what I want to emphasize here is that it is the carbon cycle operating inside stars which produces the nitrogen on which life as we know it depends. It isn't just that the elements in your body have been manufactured inside stars and scattered through space in spectacular explosions; the nitrogen nuclei in your body, in particular, have been instrumental in determining the rate at which the carbon cycle proceeded in previous generations of stars. I don't just mean nitrogen nuclei "like" the ones in your body. I mean that the very same nuclei that are now part of your body were once the dominant component of carbon-cycle reactions going on inside stars. There is a direct connection between atoms in your body and the way in which stars at least one and a half times as massive as the Sun shine.

But that is not the way the Sun shines (at least, not the main way it gets energy; a small percentage of the Sun's heat does come from the carbon cycle). After Bethe returned to Cornell from the conference in Washington, he thought long and hard about energy generation inside stars. In addition to his solo effort to work out the details of the carbon cycle, he joined forces with another physicist, Charles Critchfield, to investigate the discovery made by Atkinson a few years earlier, that the most common nuclear interaction inside a star like the Sun ought to be the simple fusion of two protons to make a deuterium nucleus, ejecting a positron and a neutrino as they did so. The chronology here is a little confusing, because the first paper by Bethe and Critchfield dealing with this proton-proton interaction was actually submitted for publication before Bethe had completed his work

2. The oxygen-16, the variety in the air that we breathe, comes into the story through the side loops, like the one I have described.

on the carbon cycle—but the carbon cycle work was completed before the proton-proton work was finished. Whatever, both breakthroughs came in the summer of 1938, and it is the breadth of Bethe's contribution to the resolution of the puzzle of how energy is generated inside stars that makes his contribution even more important than that of von Weizsäcker.

The energy generation process investigated by Bethe and Critchfield is called, for obvious reasons, the proton-proton chain (or just the pp chain). It more or less directly converts four protons into one helium-4 nucleus (plus the usual ejected positrons, neutrinos, and electromagnetic radiation in the form of gamma rays); and, as with the carbon cycle, it draws on reactions that have been studied in the laboratory and particle accelerators here on Earth and whose rates have been measured.

The pp chain really is conceptually much simpler than the carbon cycle, and the fact that it wasn't thought of first clearly highlights the way in which the imagination of the astrophysicists in the 1930s was handicapped by their unshakable conviction that heavy elements dominated the composition of the stars. As I have said, the first step in the chain is the union of two protons to form a nucleus of deuterium, which contains one proton and one neutron. As usual, the neutron has been made from a proton by ejecting a positron and a neutrino. When a second proton tunnels into the nucleus, it becomes a nucleus of helium-3 (two protons plus one neutron). Finally, when two helium-3 nuclei interact with each other, two of the protons are ejected, leaving behind a stable nucleus of helium-4 (two protons plus two neutrons). Once again, the overall effect is that four protons have been converted into one helium-4 nucleus, with two positrons, two neutrinos, and some gamma rays ejected along the way (actually *six* protons are involved in the chain, but two of them are leftover at the end).

The pp chain, which is the main source of energy in the Sun and other relatively low-mass stars.

As with the carbon cycle, there are subtleties which we need not worry about much here. The interaction between two helium-3 nuclei to make a single nucleus of helium-4 and release two protons happens about 86 percent of the time under the conditions that exist inside the Sun. Because there are traces of other light nuclei inside the Sun (notably beryllium-7), 14 percent of the time the helium-3 gets involved in other interactions, side chains which also have the net effect of converting the helium-3 to helium-4. What matters, though, is that these interactions are well understood and that the whole set of interactions accurately describes how stars like the Sun, at the lower-mass end of the main sequence, get their energy — by "burning" their nuclear fuel at a temperature of about 15 million degrees.

The modern understanding of the way in which the forces of nature operate also gives a neat insight into the longevity of stars like the

Sun. Because the ejection of positrons and neutrinos is involved, the first step in the pp chain, when two protons fuse together to make a deuterium nucleus, proceeds at a rate which depends on the strength of the weak nuclear interaction. It is the weak interaction that controls this kind of decay process, with a proton being transformed into a neutron by ejecting the other two particles. And because the weak interaction is so very weak, this interaction between two protons is very rare, over and above the difficulties involved in tunneling.[3] The other steps in the chain (unlike several steps in the carbon cycle) do not involve the weak interaction, just the strong nuclear interaction and electromagnetism, so they proceed much more quickly. Once you have made deuterium, it's all plain sailing, and stars like the Sun exist for a long time because the very first step in the chain is a bottleneck (the hole in the first bucket is tiny). One consequence of this is that any deuterium that was around when the star was born is destroyed by the action of the proton-proton chain (there is a big hole in the deuterium bucket). Overall, deuterium is not made inside stars, but is destroyed there. Which raises the interesting puzzle of where the deuterium that we detect in the atmospheres of old stars (by spectroscopy) came from in the first place. Indeed, where did the helium that, we now know, makes up 25 per cent of the mass of old stars come from in the first place? And the hydrogen itself? Before going on to tell you where and how all the heavy elements were manufactured, it is time to take a detour from the story of how stars work today, and explain where the material that the first stars were made of came from — a detour back 15 billion years in time to the Big Bang in which the Universe as we know it was born.

3. This was, in fact, taken account of when I told you how few collisions between protons result in fusion.

6

Big Bang Cookery

The history of science isn't always as neat and tidy as some of the accounts you read in books might suggest. Discoveries may come out of sequence, with notable insights that would have speeded the development of understanding sometimes not turning up for years, while on other occasions the relevance of a scientific discovery becomes clear only long after it is made. The parallel development, after about 1930, of the understanding of how stars work and of how the Universe came to be the way it is was particularly messy and confused, and although both developments depended on the new technology of improved telescopes and the new physics of quantum theory (which is why they proceeded in parallel), it took forty years for all the pieces to fit together into a self-consistent picture of how stars had evolved within an expanding Universe and where the elements that we are made of had come from. Remember that it was only at the end of the 1920s that astronomers even began to realize that stars are not made of the same stuff that the Earth is made of, and that their composition is dominated by hydrogen. At exactly the same time, Edwin Hubble and his colleague Milton Humason, working with the largest and best telescope then available on Earth, the 100-inch telescope on Mount Wilson, in California, discovered that the Universe is expanding. It was this discovery that would lead to the realization that the Universe had been born in a Big Bang, some 15 billion years ago, and that what had emerged from the Big Bang to form the first generation of stars was a mixture of roughly 75 percent hydrogen and 25 percent helium, with just a smattering of other light elements (including, crucially, deuterium). But that would not become clear until the end of the 1960s, *after* crucial developments in our understanding of how the heavier elements are made inside stars. Those developments will be described in the next chapter; but because the hydrogen and helium from which the first stars were made literally came first, from the Big Bang, it makes sense to describe that process first, even if it was not fully understood until after the theory of how stars work was completed.

I have written about the Big Bang before, and I don't intend to go into any detail about it here.[1] But I do want to emphasize that this is good science, which is thoroughly understood and has been tested by comparing theory with observations. There is little doubt that the Universe as we know it did emerge from a superhot, superdense state (the Big Bang) about (probably a bit less than) 15 billion years ago. There is, though, some discussion about how it got into that state, exactly how long ago the Big Bang happened, and also about what the ultimate fate of the Universe will be. But those debates are outside the scope of the present book.

1. In particular, see my books *In Search of the Big Bang* (London: Penguin, 1998) and *The Birth of Time* (New Haven: Yale University Press, 2000).

The "discovery" of the Big Bang began when Hubble used the 100-inch telescope to measure the distances to galaxies beyond the Milky Way. It was his work which established, once and for all, that the fuzzy blobs of light we see with our telescopes really are other galaxies. The Milky Way itself is a disk-shaped island of stars in space, about 90 thousand light years across and containing upwards of 200 billion stars. It turns out that it is almost exactly an average galaxy of its kind (although this wasn't definitely established until the 1990s). Other galaxies, called ellipticals because of their shape, are in many cases much bigger than the Milky Way; and in round numbers it is estimated that several hundred billion galaxies are, in principle, visible to our telescopes. Measuring the distances to even the nearest galaxies was a huge achievement with the technology of the 1920s, which rested upon the use of Cepheid variable stars as distance indicators. But it was the next step, taken by Hubble (who measured the distances) in collaboration with Humason (who measured redshifts), that began the story of the Big Bang.

Measuring redshifts in the light from other galaxies tells us that they are moving away. The measurements are very difficult, because although each galaxy contains hundreds of billions of stars, they are so far away that they are fainter, as viewed from Earth, than individual stars seen in our own Galaxy. But Hubble and Humason discovered not only that every galaxy they observed, except for two or three of the very nearest neighbors to the Milky Way, shows a redshift in its light, but also that the redshift is proportional to the distance of the galaxy from us. In other words, the speed with which a galaxy seems to be receding from us is proportional to its distance from us. Now, this doesn't mean that we are at the center of the Universe. This kind of redshift-distance relationship, with velocity proportional to distance, is the only kind of redshift-distance law (except for the trivial case when none of the galaxies are moving at all) which would look the same from whichever galaxy you are sitting in. It is, literally, a universal law. Everything is receding from everything else, in exactly the same way, in the expanding Universe. But why?

As soon as Hubble and Humason made the discovery, it was realized that the mathematical tools to describe what they had discovered already existed. Back in 1917, just after completing his general theory of relativity (which describes the relationship between space, time, and matter), Einstein used the equations he had discovered to try to describe the Universe at large—all of space, time, and matter. He had been baffled to discover that the equations demanded that space should either be expanding or contracting but did not allow for the existence of a static universe. In the 1920s, a very few mathematicians and astronomers tinkered with the equations, without realizing that they described the Universe we live in. But when Hubble and Humason discovered the redshift distance law (now known, rather unfairly to Humason, simply as Hubble's law), it became clear that the mathematics needed to describe what was going was already available to them.

In the 1930s, a Belgian astrophysicist, Georges Lemaître, used the combination of observation and theory to come up with the first version of what we would now call the Big Bang model of the Universe. He talked in terms of a "primordial atom" (or sometimes "primeval egg"), containing all the mass of all the galaxies in the visible Universe, sitting alone in space and then suddenly breaking apart in an explosion, like the fission of a giant radioactive nucleus. This image encouraged other people to take up the investigation of the Big Bang—but in one respect it is a misleading one, since what Einstein's equations tell us is that space itself is expanding. The Big Bang was *not* an explosion that took place somewhere in empty space, with fragments from the explosion (galaxies) flying apart through space like shrapnel from an exploding shell. Rather, what happens is that space itself expands, and takes galaxies along for the ride. It's like a piece of elastic on which you make several ink blobs. When you pull the two ends apart, the elastic expands and the blobs of ink move farther apart — but they do *not* move through the elastic.

Hard though it may be to picture, what the general theory of relativity tell us is that space and time were born, along with matter, in the precursor to the Big Bang, and that this bubble of spacetime full of matter and energy (the same thing — remember $E = mc^2$) has expanded ever since. The galaxies fill the Universe today, and the matter they contain always did fill the Universe, although obviously the pieces of matter were closer together when the Universe was smaller. Since the cosmological redshift is caused not by galaxies moving through space but by space itself expanding in between the galaxies, it is certainly not a Doppler effect, and it isn't really measuring velocity, but a kind of pseudo-velocity. Partly for historical reasons, partly for convenience, astronomers do, though, continue to refer to the "recession velocities" of distant galaxies, although no competent cosmologist ever describes the cosmological redshift as a Doppler effect.

Leaving aside the question of exactly what happened at the very beginning, when space was infinitesimally small and time had scarcely begun, for the purposes of the present book we need only worry about conditions that are thoroughly understood and have been well observed by experiments. The most extreme density of matter that we can probe today and that we thoroughly understand is the density of an atomic nucleus. Some particle physicists would argue that they understand what goes on inside nuclei, at the level of quarks. But none would deny that nuclei, and the interactions between protons, neutrons, and electrons, are so well understood as to be almost boring to a professional physicist, no matter how intriguing they may be to the layperson. Using the equations of the general theory, combined with observations of the rate at which the Universe is expanding today, we can go one better than Lemaître, winding back the expansion in our minds, and work out when it was that the density of what is now the entire visible Universe was the same as that of an atomic nucleus today. The answer turns out to be just one hundred-thousandth of a second after "the beginning." The known, and thoroughly understood, laws of physics can, in principle, describe everything that has happened since. The debate I referred to about what happened to make the Big Bang concerns that first hundred-thousandth of a second; but everything I am going to tell you about the Big Bang and primordial element production concerns the time since then, when everything is well understood.

The first person who really thought about the idea that heavy elements might be manufactured out of hydrogen in the Big Bang (the term wasn't in use then, but I shall use it for convenience and consistency) was Carl von Weizsäcker, in 1937 — intriguingly, *before* he (and Bethe) worked out how hydrogen is converted into helium inside stars! This is no real surprise, though, since at that time the received wisdom was that stars were mainly made of heavy elements, and they had to come from somewhere. Von Weizsäcker talked of heavy elements being "cooked" out of hydrogen in the early Universe, echoing Atkinson's cooking-pot analogy for stellar processes. But he also imagined this happening during an early, stationary phase of the Universe, with all the cooking being done at a suitably high temperature before the expansion began. This was just as incompatible with the equations of the general theory as the idea that the Universe might be in such a static state today.

The first person who tried to carry out quantitative calculations of what conditions might have been like in the *expanding* primordial fireball was our old friend George Gamow, in the mid-1940s. By then, people had begun to realize that the composition of stars was dominated by hydrogen (and helium) but were still worried about where the heavier elements had come from. Gamow followed up von Weizsäcker's idea that they might have been manufactured in the Big Bang, but he worked strictly within the framework allowed by the general theory of relativity, the best theory of space, time, and matter that we have. He used his skill and understanding of nuclear physics and quantum theory to calculate the kind of nuclear reactions that would have gone on in the Big Bang. He found that it wasn't as easy as he had hoped to make elements in the Big Bang. What emerges initially from the kind of conditions I have just described is a sea of protons and electrons. As long as the Big Bang stuff is still very dense and hot, some of the electrons are forced to combine with protons to make neutrons, and then some of the neutrons and protons get together, through a series of nuclear interactions, to make helium nuclei. Deuterium, by the way, is indeed manufactured in the Big Bang as an intermediate step in this process. But just when things are starting to get really interesting, the expanding fireball cools off to the point where nuclear fusion reactions stop. Gamow had found a good way to make hydrogen and helium but had failed in his attempt to explain where everything else came from. He was, however, undaunted. In the 1950s, as it became clear that stars are made up of roughly 99 percent hydrogen + helium, with only 1 percent heavy elements, the ebullient Gamow used to delight in telling his colleagues that he had explained where 99 percent of the stuff stars are made of came from, and that he was happy to leave the other 1 percent for them to tidy up.

In the course of his work on the Big Bang, Gamow also, with the help of two colleagues, made one of the most famous predictions in science, although its importance was not fully recognized for more than twenty-five years. One of the key steps in working out what went on in the Big Bang is the temperature involved. Using a combination of observations of the rate at which the Universe is expanding, the general theory of relativity, and the nuclear physics involved in making helium out of hydrogen, Gamow and his student Ralph Alpher not only calculated the temperature of the fireball but worked out that the heat radiation from the Big Bang would still be filling the Universe today. It would be quite cold, because the Universe has expanded so much; but it would now be equivalent to a sea of microwaves (like the radiation inside a microwave oven, but much colder) filling the Universe, with a temperature a few degrees above the absolute zero of temperature (0 K, which is −273 degrees on the Celsius scale).

When the time came to write a scientific paper about this work, Gamow decided that it would be a great joke to add the name of his old friend Hans Bethe as coauthor, so that it would be referred to as the "Alpher, Bethe, Gamow" paper, echoing the first three letters of the Greek alphabet (alpha, beta, gamma). Although Bethe had nothing to do with the work at all, the paper was duly published with his name on it. Gamow's delight was doubled by the fact that, entirely by coincidence, the official publication date of the journal the paper appeared in was 1 April 1948. It is indeed still referred to as the "alpha, beta, gamma" paper.

These calculations were refined by another of Gamow's students, Robert Herman, and Alpher and Herman together published another paper, also in 1948, with the more precise prediction that the Universe is filled with a sea of microwave background radiation at a temperature of about 5 K. But they were ahead of their time. Nobody took the calculation seriously enough to search for this radiation, and no real progress was made in the theory of what went on in the Big

Bang until the 1960s. Indeed, in the 1950s there was an intense debate between two schools of cosmological thought about whether there had actually been a Big Bang.

The debate was stimulated by three researchers, Herman Bondi, Tommy Gold, and Fred Hoyle, who pointed out that the observed expansion of the Universe could also be explained if, instead of everything having been "created" together in a Big Bang, there could be a process of "continual creation," with new atoms (presumably of hydrogen) appearing out of nothing at all in the wide open spaces between galaxies as the galaxies moved apart.[2] These new atoms would eventually get together to form new galaxies, so that overall the Universe always looked the same, even though the galaxies visible at any one epoch are moving apart. This is not as wacky an idea as it is sometimes made to seem today — if stresses in spacetime can "create" the mass of the Universe in one go, why shouldn't similar, but smaller, stresses in spacetime associated with a steady expansion of the Universe create atoms one at a time? Like all good scientific ideas, the steady state model, as it became known, was testable. It predicted that the way the Universe is expanding should look exactly the same for very distant galaxies as it does for nearby galaxies. The Big Bang model, of course, implies that the Universe is changing as it ages. Since we see distant galaxies by light (or other radiation) which left them long ago, when the Universe was younger (in some cases, the light has been billions of years on its journey), this aging of the Universe should show up if we compare the properties of nearby galaxies with those of very distant galaxies. A huge observational effort went into testing the rival ideas in the 1950s and early 1960s, mainly

2. The use of the term "created" by cosmologists does not imply the existence of a creator, any more than their equivalent use of the term "born" implies the existence of a mother. Both are just shorthand terms to refer to the beginning of the Universe as we know it.

from radio astronomers, who had the technology to probe farther out into the Universe at that time than their optical astronomer counterparts. They also had the incentive to do this — there was some bad feeling between Hoyle and the radio astronomers in Cambridge, where Hoyle worked, and this encouraged the radio astronomers to try their damnedest to prove the model he supported was wrong. The feeling was mutual. Although it was, in fact, Hoyle who coined the term Big Bang in its cosmological context, in a BBC radio broadcast in 1950, the kindest thing he ever said about the model was that it was "inelegant."

Inelegant or not, the eventual outcome of the observational tests was a decisive win for the evolving Universe model. But by the time the jury reached its verdict, compelling evidence in favor of the Big Bang had already come in, from the entirely accidental discovery of the background radiation that had been predicted by Gamow and his colleagues back in the 1940s.

The discovery was made by two young researchers, Arno Penzias and Robert Wilson, working with a radio telescope owned by the Bell Laboratories and originally designed for experimental work with communications satellites. Before they could begin using the instrument for radio astronomy, Penzias and Wilson had to make sure they understood all its foibles and to calibrate it by looking at known sources of radiation. They also pointed it at the empty sky, between known radio sources, to check the zero reading of the instrument. To their frustration, they found that it was plagued by a persistent radio noise, rather like the static you hear when an AM radio is tuned off-station, which seemed to be coming from all directions in the sky and which they assumed must be a fault in the antenna or its amplifier system. Either that, or the Universe was filled with microwave radiation with a temperature of a few K — a notion they dismissed as ridiculous.

Unknown to the Bell Labs team, just fifty kilometers away from their base in Holmdel, New Jersey, a team of astronomers at Princeton University, headed by Jim Peebles, was building a detector specifically designed to find the background radiation — not because of the work by Gamow and his colleagues, which had long since been forgotten, but because Peebles had made essentially the same calculation, independently of the earlier work. In December 1964, Penzias mentioned the problem he and Wilson were having with excess noise in their radio telescope to a colleague at the Massachusetts Institute of Technology; in January 1965, that colleague telephoned Penzias to say that he had just heard about a talk Peebles had given in which he predicted that the Universe should be filled with a sea of microwave radiation with a temperature less than 10 K. The two teams quickly got together, and Peebles confirmed that Penzias and Wilson had found the radiation he had been about to start looking for.

When the discovery was officially announced, in a pair of papers published later in 1965 (one by Penzias and Wilson, one by the Princeton team), people began to take the idea that there really had been a Big Bang seriously. Before, cosmology had been almost like an intellectual game, something for mathematicians to do with the equations of the general theory of relativity, but never as real as the study of things like stars, which we can see with our eyes. The only real fire in cosmological debates had been provided by arguments driven by personal rivalries or animosity, like the desire to prove Hoyle wrong, which drove the Cambridge radio astronomers on to measure the properties of very distant radio galaxies as accurately as possible. Indeed, even in 1965, when I had the opportunity, as an undergraduate at the University of Sussex, to discuss my career prospects with Herman Bondi himself, he advised me to give up any thoughts of working in cosmology (then my main ambition), because it would be a dead-end career. But his advice was being overtaken by events. The background radiation soon began to put things in a different perspective. It might not be tangible to our human senses in quite the way that starlight is, but it was certainly tangible to our electronic detectors, and every bit as tangible as the radio noise from distant galaxies that was proving so useful in the debate about the nature of the Universe. The discovery of the background radiation (which actually turns out to have a temperature of just under 3 K) forced astronomers to take seriously the idea that there really was a Big Bang; and it also encouraged physicists, who had previously regarded cosmology as a rather effete subject, more philosophy than real science, to begin to take the notion of the early Universe seriously.

But Big Bang cosmology, as a fully respectable branch of physics, did not achieve that status overnight, even with the discovery of the background radiation. Those astronomers and physicists who had already been interested in the nature of the Big Bang (such as Peebles) were, of course, readily persuaded that what had been found really was the "echo of the Big Bang." But other researchers looked to see if there might be alternative explanations — even Penzias and Wilson weren't at all sure at first what it was that they had found (as it happens, they both rather liked the steady state model), and their discovery paper simply reported what they had observed, with no interpretation of the data, though they referred to the accompanying paper by Peebles and his colleagues for "a possible explanation for the observed excess noise temperature." The piece of work which finally convinced many physicists that the Big Bang cosmology was serious quantitative science, not airy-fairy wishful thinking, was published two years later, in 1967; it dealt with the way light elements were cooked in the Big Bang. Crucially, though, one step in this calculation used the new observations of the temperature of the background radiation today to give a guide to exactly what the conditions were like in the Big Bang itself.

One of the most intriguing things about this seminal work on Big Bang nucleosynthesis, as it is called, is that Fred Hoyle was a key member of the team that carried out the work—the same Fred Hoyle that was the leading proponent of the steady state cosmology. So what was he doing working on the theory of the Big Bang? He was simply being a good scientist. The fact that he had a personal preference for the rival model of the Universe didn't mean that he couldn't use his skills as a physicist and mathematician to work out what would have happened under the conditions that existed in the Big Bang fireball *if* there had been a Big Bang fireball.

This neatly makes a point about the way science progresses. There has to be an element of speculation—a guess, based on past observations and experiments, combined with intuition about the way the world works. "What if," you might imagine Isaac Newton musing, "gravity obeys an inverse square law?" Then, you test the guess, by using it to make predictions which you can compare with the outcome of experiments and observation of the way the real world works. You don't have to *believe* in what you are testing, the way people believe in religion. You make a guess (or somebody else makes a guess) and you test it. Hoyle guessed that the simple steady state model was a good description of the Universe; somebody tested it, and found that he was wrong.[3] Other people guessed that the Universe was born in a Big Bang. Hoyle tested that guess, and found

3. I use the term "simple" here advisedly. To this day, Hoyle still espouses a version of the old steady state model which is much more complicated, and would include what we call the Big Bang within itself. This is essentially the same as what is usually called "inflationary cosmology"; but that story lies outside the scope of the present book.

compelling evidence that they were right. Indeed, in many ways it is much more impressive that the test had been carried out by a skeptic, since you know he wasn't deluding himself with wishful thinking in his interpretation of the results. In a similar way, in the early decades of the twentieth century, Albert Einstein guessed that light behaved like a stream of tiny particles, now called photons. The American experimenter Robert Millikan was infuriated by this, and spent ten years trying to prove Einstein wrong. He succeeded only in proving that Einstein's guess was a good one. This is much more persuasive, to an outsider, than if Einstein had done the experiments and claimed that he had proved his own guess was right!

Hoyle's ability to focus on one problem at a time made a deep impression on me when I was a student at the Institute of Astronomy in Cambridge, where Hoyle was director, in the late 1960s. Hoyle used to say that he liked to "compartmentalize" his research, so that prejudices from one part of his work wouldn't influence what he was doing in another area; and he also always carefully avoided getting involved in carrying out astronomical observations, because he felt that his theoretical prejudices might unconsciously influence his data gathering, so that he might miss something important that didn't fit his preconceived ideas. He always felt that observers should do the observing, without prejudice, and that theorists should try to explain the honestly obtained observational data. Interestingly, Hubble felt the same way. He always concentrated on the observations and didn't try to explain them in theoretical terms, leaving that to the theorists.

Hoyle's contribution to the breakthrough paper on Big Bang nucleosynthesis, the paper that was (ironically, in view of his own prejudices) to be regarded as sounding the death knell of the steady state model, stemmed from work he had been carrying out in the 1950s on the way elements are formed inside stars — stellar nucleosynthesis. This work is the subject of the next chapter; all we need to know here is that by the early 1960s it was clear that although the heavy elements really could be manufactured inside stars, there was no way in which the huge amounts of helium (about 25–30 percent) then being measured in stars by spectroscopy could all have been manufactured out of hydrogen, the simplest element, inside the stars themselves. Stellar nucleosynthesis could be responsible for only about a tenth of the helium seen in stars. The rest had to come from somewhere else — from the primordial material that stars were made out of in the first place.

Hoyle was an old friend and sparring partner of George Gamow, and in the 1950s they often used to try (in a friendly fashion) to persuade each other that the other's cherished cosmological ideas (Gamow's Big Bang; Hoyle's steady state) were wrong. As a result, Hoyle knew a great deal about Gamow's work and was one of the few astronomers who were still aware, in the early 1960s, of the prediction of the existence of a sea of cosmic-microwave background radiation — although even Hoyle didn't realize how easy it would be to find this radiation. But things had moved on since the 1940s: there was a better understanding of the theoretical side of nuclear physics and better experimental determinations of the rates at which nuclear interactions important under Big Bang conditions go. While preparing a lecture course on cosmology, Hoyle, in Cambridge for the academic year 1963–64, decided that the helium problem was so severe that it justified reworking the calculation made by Gamow's team to take account of the latest nuclear physics data. This work was carried out by Hoyle and his colleague Roger Tayler, and the results appeared in a paper published in 1964. Like Gamow's team, they concluded that *if* there was a Big Bang, roughly the right amount of helium would be produced if matter had been processed in a high-temperature fire-ball. In their own (cautious!) words, the results of their calculations "could be interpreted as evidence that the Universe did have a singular origin." According to the new calculations by Hoyle and Tayler, the fireball was so energetic that there were about a billion photons for every nucleon (every proton or neutron). The numbers were not exact, because Hoyle and Tayler had only a rough idea of how much primordial helium they needed to be produced in the fireball. But the billion photons per nucleon emerging from the Big Bang were the radiation that would form the cosmic background radiation detectable today. As part of this work, Tayler calculated the implied temperature of the background radiation in the Universe today, assuming that a mixture of about 25 percent helium and 75 percent hydrogen, matching spectroscopic observations of the oldest stars, had emerged from the Big Bang. For once, though, Hoyle did let his prejudices get the better of him, and, as Tayler ruefully recalled later in life, this aspect of their work was downplayed in the published version of the paper.

The paper by Hoyle and Tayler stimulated interest in Big Bang nucleosynthesis, and this interest was spurred even more by the discovery of the background radiation, announced the following year. Hoyle himself developed the work further with his friend Willy Fowler (an expert on nuclear physics), at the California Institute of Technology, and with Robert Wagoner, a student of Fowler's. In 1967, the team of Wagoner, Fowler, and Hoyle published the results of their much more detailed calculation of Big Bang nucleosynthesis, matching not only the calculated helium abundances but also the abundances of the light elements lithium and deuterium to the composition of the oldest stars, with everything calibrated against the measured temperature of the background radiation, 2.76 K. It wasn't just that the nuclear physics calculation made the Big Bang respectable; it was the match with the background radiation that made the interpretation of the background radiation as a relic of the Big Bang itself respectable and convinced the doubters that it really was the echo of the fireball of creation. It was the moment when Big Bang cosmology came of age.

This was an absolutely stunning achievement, which made a deep impression on me at the start of my career in astronomy. In the autumn of 1966, I started an M.Sc. course in astronomy at the University of Sussex. The first major astronomical talk I attended as a student was one given by Wagoner in Cambridge (it was also my first visit to Cambridge), during which he described this work, before the paper had even been published. It was clear, even to a new student, that this was the moment when Big Bang cosmology became respectable science, involving comparison between numbers predicted by theory and numbers that could be measured by experiments in nuclear physics labs or observed by studying the compositions of stars. There was a tremendous buzz associated with knowing that not only was this a major landmark in science but that for the time being I was one of only a few people in the whole world—maybe a couple of hundred—who knew about it.

Of course, the calculations (and the observations and experiments) have been refined further since 1967. But the picture is still essentially the same, and what matters for an understanding of where the elements that we are made of came from is that what emerged from the Big Bang was a mixture of about 75 percent hydrogen, just under 25 percent helium, and a smattering (but a smattering that can be calculated quite accurately) of very light elements such as deuterium and lithium. But there was no trace at all of the heavier elements that we are interested in (and made of)—carbon, oxygen, nitrogen, and the rest — having been manufactured in the Big Bang. Not a single atom, according to these calculations. The big question then is, how was the very light primordial stuff turned into the stuff we are made of? And that question had already been answered in the 1950s, before the Big Bang model became firmly established.

7

Burbidge, Burbidge, Fowler, and Hoyle

Fred Hoyle's long involvement with the search for the origin of the chemical elements began in 1944, when he was twenty-nine years old and working on radar as part of the British war effort. That year the job took him on a liaison mission to the United States and Canada. While in the Los Angeles region on official business, he found time on a free weekend to visit the astronomers at the Mount Wilson Observatory, where a discussion with Walter Baade roused his interest in great stellar explosions, novae, and (in particular) supernovae. A little later, on the Canadian leg of his journey, Hoyle met up with members of a team of British nuclear physicists, based near Montreal. They were ostensibly there to work with the team in Chicago who were building the world's first nuclear reactor — but were really, as Hoyle describes it, a "listening post" trying to pick up details of the Manhattan Project (the project to build the first atomic bomb), which the Americans wanted to keep secret even from their closest allies. There,

Hoyle himself picked up enough information about the atomic bomb project to begin to wonder if a supernova explosion might operate in the same way as the technique that he surmised was being used by the bomb makers (nobody told him; he worked it out from what people were *not* talking about) — with an initial implosion forcing material together at extreme temperatures and pressures, thereby triggering an even larger explosion as the material turned inside out and blew itself to bits. Back in England, Hoyle thought about the problem in his spare time in the winter of 1944–45. He knew that nuclear fusion occurred because the energy balance favored the production of heavier elements out of light ones, but only up to a point. The arrangement of protons and neutrons as helium nuclei is more desirable, in energy terms, than if the same number of particles are arranged as hydrogen nuclei; carbon nuclei are even more energetically desirable than helium nuclei, and so on. But there is still the old problem of getting positively charged particles such as protons or alpha particles to move fast enough to penetrate the field of electrical repulsion around a nucleus in order for fusion to occur. Even with the help of the tunnel effect, it is much harder for another positively charged particle to penetrate a nucleus which contains more protons, because there is more electrical charge to overcome. So each step in the process needs a hotter cooking pot, even if the end product represents a more energetically desirable nuclear state.

In these terms, the most desirable nuclei are those of elements such as nickel and iron. Making heavier nuclei than this requires an input of energy over and above the energy needed to overcome the electrical repulsion. This could be achieved if a star made largely of nuclei like carbon and oxygen collapsed in upon itself (the implosion), releasing so much gravitational energy that as well as forming a large amount of things like iron and a smattering of the heavier elements, many

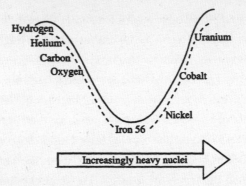

The "valley of stability." The most stable nuclei are those in the middle range of masses, centered around iron-56. Light nuclei (to the left) would "like" to fuse together to make heavier nuclei, releasing energy and rolling down the valley as they do so. But to make nuclei heavier than iron-56, energy has to be provided to make the nuclei stick together, pushing them up the other side of the valley.

nuclei would be smashed to bits, great floods of protons and neutrons would be released, and the star would blow itself apart. Could this, Hoyle wondered, be what happened in a supernova?

When Hoyle carried the calculation through in the spring of 1945 (as best he could under the difficult wartime circumstances) and worked out the kind of conditions that would be required to produce the kind of proportions of elements seen on Earth, he found that the stars in which iron was cooked would have to reach the staggering temperature of 5 billion degrees, which is enormous when compared with the temperatures at the hearts of main-sequence stars, a mere 15 million to 20 million degrees. So, Hoyle reasoned, since we know iron exists in the Universe, all the intermediate temperatures between 20 million degrees and 5 billion degrees must occur somewhere inside stars during the course of their evolution, and the nuclear reactions associated with this vast range of temperatures could certainly produce all the elements in just the observed proportions. At least, he hoped so. The details were still distinctly hazy, but Hoyle already had the outline clear when he left his work on radar and returned to academic life in Cambridge in the summer of 1945. His first paper on the origin of the elements, in which he discussed these ideas, was published a year later, in 1946 — the same year that Gamow and his students were starting to suggest that all the chemical elements could have been cooked up in one go, in the fireball of the Big Bang.

The key feature of Gamow's model (apart from the fact that everything is thought to have happened in the Big Bang) is that it involves the heavier elements being built up from hydrogen steadily, by the addition of neutrons to existing nuclei. The first steps are rather like (but not exactly the same as) the proton-proton chain that (we now know) operates inside the Sun. A proton captures a neutron to form a deuteron. Some deuterons then each capture a neutron to form tritium, an unstable nucleus of hydrogen-3 which promptly decays, spitting out an electron (the process known as beta decay) as one of the neutrons is converted into a proton. The nucleus has become one of helium-3, and can capture another neutron to form helium-4 (an alpha particle), and so on. Neutron capture and beta decay are all you need, said Gamow, to make all the elements.

As well as being simple, this idea had two things going for it in the 1940s. First, experiments had shown that virtually all nuclei did indeed absorb neutrons if they were bombarded with them. Even better, slightly more sophisticated experiments had shown that the reaction rates for this process for different nuclei (the "neutron-capture cross-sections," in technical terminology) lead to a prediction for the relative abundances of the elements which agrees pretty well with the observed abundances — indeed, this match between theory and observation was one of the linchpins of Hoyle's version of nucleo-synthesis (taking place inside stars, not in the Big Bang, but still involving neutron captures). Nuclei that are good at capturing neutrons ought to be rare, because they are quickly converted into other elements; nuclei that are bad at capturing neutrons ought to be relatively common, because they form bottlenecks in the process. These effects can be quantified, which is where the reasonably good agreement between experiment and observation comes in.

But although the ebullient Gamow tended to dismiss the difficulties of making anything heavier than helium as a minor detail, even Ralph Alpher and Robert Herman themselves were, within a few years of the 1946 paper, drawing attention to two key difficulties with the model. The first is worrisome, but did not, at the time, necessarily seem insurmountable. The Big Bang happened very quickly. In round terms, as Steven Weinberg memorably made famous in his best-selling book, the conditions under which nucleosynthesis could go on in the Big Bang lasted for just over three minutes. Would that really have been long enough for all the neutron capturing and beta decaying that would have been needed to make the variety and abundance of chemical elements in the Universe today? But this problem fades into insignificance compared with the other, literally insurmountable difficulty with the idea that all the elements were made solely by neutron capture and beta decay, either in the Big Bang or anywhere else. There is no stable nucleus which contains a total of five nucleons, nor is there a stable nucleus which contains eight nucleons. So there are two gaps in the ladder of atomic masses, right at the start of the process. It is possible, in experiments on Earth, to make helium-5 by bombarding helium-4 with neutrons. But it promptly spits out the extra neutron, reverting to helium-4—*very* promptly, before it could absorb any other neutrons. Similarly, it is possible to make beryllium-8 artificially, but it almost immediately splits into two helium-4 nuclei. There is no natural helium-5, and there is no natural beryllium-8. If neutron capture were the only way to make heavier elements than helium, it would be impossible for nature to manufacture these elements, either in the Big Bang or inside stars. Something else is required, in addition to neutron capture, *wherever* these elements were made.

The way Hoyle solved the problem of these mass gaps was the key to the breakthrough which led, in the 1950s, to a full understanding of how the chemical elements are manufactured inside stars. The great thing about stars as factories for manufacturing the elements is that they live so long. If everything happened in three minutes, all the processes involved would have to be pretty efficient. But in a star that lives for millions — even billions — of years, there is plenty of time for even rare events to do their bit in the overall manufacturing process. In 1951, two astronomers, Ernst Öpik and Edwin Salpeter, each independently suggested a way to jump across both the mass gaps in one go, using a minor, rare nuclear interaction. They reasoned that if three nuclei of helium-4 (three alpha particles) collided with one another simultaneously inside a star, then they could fuse to form a single nucleus of carbon-12, without the bother of having to make either helium-5 or beryllium-8 along the way.

The problem was, even over the lifetime of a star this rare interaction would not be able to build up a significant amount of carbon. In fact, there were two problems — first, the triple-alpha process, as it became known, was so rare that it would make only a tiny amount of carbon. Second, carbon-12 itself reacts rather eagerly with helium-4, absorbing an alpha particle and becoming a nucleus of oxygen-16. So what little carbon was manufactured by the triple-alpha process ought all to have been turned into oxygen about as soon as it was made. But we know there is a lot of carbon in the Universe. So Hoyle reasoned that there must be something special about the triple-alpha process which makes it occur much more frequently (more efficiently) than seems likely at first sight.

The best way to picture what is going on is not as a precisely simultaneous collision between three alpha particles in the heart of a star, but as a two-stage process. Alpha particles (helium-4 nuclei) must be colliding with one another quite often under those conditions, forming beryllium-8 nuclei which quickly split apart.[1] But because new beryllium-8 nuclei are being produced all the time, there is always a certain proportion of them present — about one nucleus in every 10 billion in a star with a central temperature of about 100 million degrees will be a nucleus of beryllium-8. None of these beryllium-8 nuclei exist for more than a split second, but as fast as they break apart they are replaced by new nuclei of beryllium-8. So Hoyle looked for a reason why the process by which these rare beryllium-8 nuclei latched on to another alpha particle might be so efficient that it turned many of them into carbon nuclei before they could decay.

Everything came together at the beginning of 1953. Hoyle had been invited to spend the first three months of that year at Caltech, where he was to give a series of lectures on nucleosynthesis. It was in preparation for this that he took a long hard look at the problem of how carbon could be made inside stars, and decided that it was only possible if the carbon-12 nucleus itself could exist in what is known as an excited state, or resonance. In itself, this was no surprise — all nu-

---

1. How quickly? The lifetime of a beryllium-8 nucleus is about 10^{-19} seconds, a decimal point followed by 18 zeros and a 1, which is a "split second" by anyone's definition!

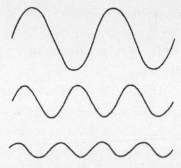

The resonance which exists in an excited nucleus of carbon-12 can be likened to the way different musical notes can be played on a single guitar string. Each note corresponds to a different harmonic of the fundamental note for the string, but the different notes (with different wavelengths) must all fit the length of the string with the ends held still. In an analogous way, the excited state of carbon-12 can be thought of as a high note played on the basic carbon-12 string.

clei can exist in excited states. But what was special about Hoyle's idea was that it required the energy of the carbon-12 resonance to have a precise value — and no experiment, as far as he was aware, had ever measured a carbon resonance with this energy.

You can think of these resonances as rather like the higher harmonics that can be played on a plucked guitar string. The open string makes a particular note (equivalent to what is called the ground state of the nucleus), but it can also vibrate at a set of higher harmonics, which are determined by the length of the string (equivalent to the resonances for a particular kind of nucleus). If there was no resonance at the energy level Hoyle calculated, then you would expect a fast-moving alpha particle colliding with an individual beryllium-8 nucleus to blow it apart. But if there was a resonance in the carbon-12 nucleus at just the right energy, then the incoming alpha particle could gently slot into place in the nucleus, forming an excited state of carbon-12 which could then radiate energy away (in the form of a gamma ray) and settle down into an ordinary, unexcited nucleus of carbon, in its ground state.

This happens even though the carbon-12 nucleus doesn't yet exist in order to resonate. The act of an alpha particle colliding with a beryllium-8 nucleus creates an excited form of carbon-12. In an analogous way, the musical note doesn't exist until the guitar string is plucked — it has the potential to exist, but it is created by the act of plucking the string. In the case of the carbon-12 resonance, the astrophysical requirements were so precise that Hoyle was able to calculate the exact energy required for the resonance, 7.65 million electron Volts (MeV) above the ground state of carbon-12. If the energy level were even 5 percent bigger than this, the process would not work. He took the idea to Willy Fowler, the nuclear physicist based at Caltech whom we have already met, and asked if it would be possible to check whether carbon-12 did possess the required resonance.

As Fowler used to tell the tale, he thought that Hoyle was crazy, but when the visitor from England insisted, Fowler put together a small team to carry out the necessary experiment, as much to shut Fred up as in any expectation of proving him right.[2] Of course, they did prove him right — one of the most stunning examples of a theory making a prediction which was tested by experiment and proved correct in the entire history of science. From the simple fact that carbon exists, and a basic understanding of how hot stars are inside, Hoyle had predicted the value of a seemingly esoteric property of the carbon nucleus to an accuracy better than 5 percent. With the help

2. Hoyle's memory differs on this point; at the very least, he says, Fowler never openly expressed any doubts about Hoyle's sanity.

of Fowler and his colleagues, Hoyle had jumped the mass gap and shown how stars could manufacture elements heavier than helium.[3]

But all this rests on a remarkable triple coincidence, which is worth spelling out before we look at the detail of how heavier elements are made. First, it would actually be a bad thing (speaking as a human being) if beryllium-8 were stable, or even if it were slightly less unstable than it is. If the process which converted two alpha particles into a single beryllium-8 nucleus were much more efficient than it is, once stars had used up all the hydrogen fuel in their hearts, helium would suddenly be converted into beryllium, releasing so much energy that the star would be blown apart, and no heavy elements would ever be made. Second, the carbon-12 resonance is, as we have seen, just at the right level for an incoming alpha particle to slide gently into place as it links up with a beryllium-8 nucleus. If it were a littler higher, then the triple-alpha process would be so rare that only tiny amounts of carbon would be made, they would all be turned into oxygen, and there would be no carbon in the Universe. Third, it happens that there is a similar resonance involving oxygen-16. It has an excited state at an energy 7.19 MeV above the lowest energy state of this particular nucleus. But the amount of energy that can be provided when an alpha particle interacts with a nucleus of carbon-12 under the conditions that exist inside stars is 7.12 MeV. If these two numbers were the other way around, or if the oxygen-16 resonance energy were a mere 1 percent lower than it actually is, all the carbon manufactured inside stars would be rapidly turned into oxygen-16. Once again, in that

3. It's taking nothing away from the achievements of Hoyle and Fowler to mention, for the sake of scrupulous historical accuracy, that one experiment had hinted at the existence of an excited level of carbon-12 in the 1930s, but other experiments failed to confirm it, so by the time Hoyle came on the scene the claim had been forgotten, and he genuinely predicted what Fowler would find.

situation there would certainly be no carbon-based life forms sitting around puzzling over the origin of the elements. All this may be just coincidence, or it may be telling us a deep truth about the way the Universe works—a point I shall return to later. For now, though, with the mass gap bridged it is time to complete the story of stellar nucleosynthesis.

There are two aspects to the story. The first is working out the nuclear physics involved—things like neutron-capture cross-sections, and the way in which alpha particles are added to nuclei (now known as the alpha process). The second is explaining how the conditions required for these processes to operate occur inside stars. And I shall take them in that order, although the work was actually done with both strands being followed in parallel, with progress in one strand stimulating progress in the other and feeding back into the work on the first strand, and so on.

Hoyle's first major paper on the alpha process and the associated nuclear interactions that explain how all the elements from carbon to nickel are synthesized in stars appeared in 1954. But this was still a preliminary attempt at explaining stellar nucleosynthesis, with many details about the way the process could produce the exact abundances of the elements observed in nature still to be resolved. After Hoyle returned to England, a full teaching load in Cambridge and commitments to other research projects meant that he had to abandon the nucleosynthesis work for a while. But it happened that the following year Fowler visited Cambridge on a sabbatical from his post at Caltech. There, he got to know the British husband-and-wife team Geoffrey and Margaret Burbidge, who were puzzling over the exact abundances of different elements in stars that were being revealed by new advances in stellar spectroscopy.

Margaret Burbidge had always been an astronomer, but Geoffrey

Burbidge was a physicist who got sucked into astronomy through working with Margaret; in spite of his work testing Hoyle's carbon-resonance prediction, in 1954 Fowler was also very much a nuclear physicist—but he was about to be sucked into astrophysics in a big way as well. The Burbidges and Fowler tackled the problem of explaining the element abundances in stars, and decided that many of the observed features could indeed be explained if there were a steady supply of neutrons which could be absorbed by nuclei in much the same way that Gamow had envisaged, one at a time with radioactive decays in between. This became known as the s-process, where the s stands for slow. Hoyle kept in touch with this work, and would have been keen to join in the calculations, had his other commitments allowed. But he was able to pass on early news of a breakthrough that had been made on the other side of the Atlantic, by the Canadian nuclear physicist Alastair Cameron.

The obvious question to ask about neutron capture inside stars is, where do the neutrons come from? Left to their own devices, free neutrons decay into protons and electrons (plus neutrinos) in a few minutes—no problem if you are trying to make all the elements in the three minutes allowed by the Big Bang, but a serious difficulty if you want to spend billions of years making them inside stars. Cameron wrote a paper suggesting that neutrons might be supplied inside stars by a reaction involving the isotope carbon-13. When carbon-13 absorbs an alpha particle, it is transformed into a nucleus of oxygen-16, spitting out a neutron as it does so. Cameron's suggestion was premature in one sense, because at the time he could not explain how the carbon-13 was manufactured inside stars. As a result, when he submitted the paper to the *Astrophysical Journal*, two referees recommended that it be rejected. The editor of the journal, Subrahmanyan Chandrasekhar, wasn't so sure. He took the unusual step of asking for a third opinion. The person he asked was Hoyle, who saw Cameron's suggestion as a profound insight and recommended publication. The paper appeared in print in 1955, and Cameron followed it up with a great deal of work essentially duplicating, independently, some of the work carried out by Hoyle, Fowler, and the Burbidges over the next few years, especially the calculations of the neutron-capture processes. Since then, astrophysicists have also identified the nuclear interactions that go on inside some stars late in their lifetimes and do indeed make the all-important carbon-13. But although Cameron deserves full credit for his major contribution to the understanding of stellar nucleosynthesis,[4] the complete, definitive package was indeed put together by Hoyle, Fowler, and the Burbidges, when the entire team assembled in California in 1956 — Fowler back at his home base of Caltech, Hoyle on another short-term visit from England, and the Burbidges on what became essentially a permanent move, triggered by their work with Fowler in Cambridge.

The work that the team carried out in California in 1956 was very much a team effort, and when the fruit of their labors was published as a huge scientific paper in 1957, in the October issue of *Reviews of Modern Physics,* the names of the authors were listed, democratically, in alphabetical order, as Burbidge, Burbidge, Fowler, and Hoyle. It became known, from the initials, as B[2]FH, and is still a legendary landmark in physics, not just astrophysics.[5] I well remember the first time I held a copy of the paper in my hands, in the autumn of 1966, and understood that the work it described explained the origin of the

4. He might have received more credit at the time, except that he was working for Atomic Energy Canada Limited, and at that time of Cold War secrecy his most important results were first published in a classified report that few people had clearance to read.

5. The democracy of the alphabetical arrangement of the names belies the actual order in which they joined the project: Hoyle started it all, he enlisted Fowler, and Fowler brought the Burbidges on board.

chemical elements, including the elements in my own body; it was, literally, a spine-tingling moment. If ever a piece of physics was obviously worthy of a Nobel Prize, the B²FH work was it. But thereby hangs a tale.

The Nobel Foundation is hamstrung by many curious rules and rituals, and also, on occasion, by stupidity. One bizarre rule is that no more than three people can share the prize for one piece of work, which would be inconvenient even to the best-intentioned judge trying to acknowledge this particular study. But everybody outside Stockholm—including Willy Fowler—was totally astonished when, in 1983, the prize awarded for the work on stellar nucleosynthesis went to Fowler himself, alone out of the team. As Geoffrey Burbidge put it in an obituary of Fowler published in the *Quarterly Journal of the Royal Astronomical Society,* "the singling out of Fowler caused some strain among B²FH, since we were all aware that it was a team effort and the original work was done by Fred Hoyle." If one person had to be singled out for the award by the rules of the Foundation, certainly that person ought to have been Hoyle. Why was he overlooked?

There are two possible reasons, both of which seem plausible in view of the Nobel Committee's track record. Hoyle's own belief is that it was a punishment because he had had the temerity to criticize the award of the 1974 Nobel Prize for physics to the radio astronomer Antony Hewish, specifically for the discovery of pulsars. It happens that pulsars were actually discovered by Jocelyn Bell, who at the time was a graduate student working under the supervision of Hewish. Although Hewish picked up the discovery and worked on the pulsar problem, Hoyle was not the only person saying, in 1974, that it was rather odd that the person who made the discovery hadn't gotten a share of the prize. But he may have suffered the consequences of this criticism nine years later, when Fowler received his award. The other possible objection that the Nobel Committee may have had to giving Hoyle the prize is that in the 1970s he had become interested in the origin of life, and had published several papers suggesting that life might have evolved in space and that diseases such as influenza might be brought to Earth by comets. These ideas are often regarded as cranky, and are certainly not mainstream, although they are based on hypotheses which are at the very least worth testing. It may be that the Nobel Committee didn't want to lend respectability to what they saw as cranky ideas by giving Hoyle the prize — if so, a clear lesson is to be drawn from the example of Francis Crick, who shared the Nobel Prize with James Watson for their discovery of the structure of DNA, but who published his unconventional ideas about life and the Universe, not just panspermia but elaborations on that theme echoing some of Hoyle's ideas, later in his life, with the prize safely under his belt.

But science isn't about prizes and awards, and none of this has much bearing on the story of where we come from. The remarkable thing about B²FH is that it explains the relative abundances of the elements in great detail, using accurate calculations involving neutron-capture cross-sections, the alpha process, and so on. Obviously, I am only going to sketch the outlines of how they did this here, but it is important to appreciate that this was not some vague speculation, the way we might imagine Gamow saying, "Well, if hydrogen and helium emerge from the Big Bang, everything else must be done inside stars." It told us, very precisely, just how everything else (in fact, carbon and everything heavier than carbon) is made inside stars.

There are three key steps in the process, plus some important foundation work in which Fowler produced the definitive study of all the processes by which hydrogen can be converted into helium inside stars, and thereby established, once and for all, that no more than 20 percent of the helium in the Universe (probably not even that) could have been made in this way. The first is the alpha process, and variations on the theme, in which energy is generated inside stars by the addition of alpha particles to preexisting nuclei. Three alpha particles make a nucleus of carbon-12; adding one more gives you oxygen-16; adding another gives neon-20, and so on. Even without doing detailed calculations, it is clear that this must be a fundamental series of interactions going on inside stars, because the most common elements (apart from hydrogen and helium) are just those produced by the alpha process — oxygen, carbon, nitrogen, silicon, magnesium, neon, and iron (I describe how this happens in more detail in the next chapter). A key feature of this process, as I have mentioned, is that each step has to take place at a higher temperature, because the heavier nuclei contain more protons, so they have a larger positive charge, which repels the positively charged alpha particles that approach them more strongly. Even with the aid of the tunnel effect, an alpha particle has to move faster to penetrate an oxygen nucleus, containing 8 protons, than it does to penetrate a carbon nucleus, containing 6 protons.

All the time these processes are going on, nuclei also have a chance (many chances!) to absorb a neutron by the s-process, to decay, and then to absorb another neutron, and so on (the neutrons that drive the s-process do indeed come from carbon-13, as Cameron suggested, but may also be contributed by similar interactions involving alpha capture by the isotopes oxygen-17 and neon-21). This is how many of the elements which do not have nuclei made up of whole numbers of alpha particles are manufactured inside stars. But all this effectively stops with nuclei of iron-56 and nickel-56. Iron-56 nuclei contain 26 protons and 30 neutrons, while nickel-56 nuclei contain 28 protons and 28 neutrons, effectively 14 alpha particles fused together. Up to that point, fusion of lighter nuclei into heavier nuclei releases energy (see figure on p. 132). But fusion of anything with nuclei of iron and nickel (or heavier nuclei) requires an input of energy, essentially because there is now so much positive charge on the nucleus (eventually, for nuclei heavier than uranium, the positive charge overwhelms the strong nuclear force and blows everything apart).

Because many elements heavier than iron and nickel do exist in nature (even though only in relatively small quantities), Hoyle and his colleagues knew from the outset that there must be a natural process which built them up inside stars. The obvious way to do this is with neutrons, which do not have to worry about the positive charge on the nucleus they are approaching. But if the neutron captures proceeded slowly, these heavy elements would, in many cases, decay before they could capture enough neutrons to build up really heavy, stable nuclei of things like gold and lead. There had to be a process of rapid neutron capture (called, naturally enough, the r-process), in which a flood of neutrons swept through the nuclei, so that an individual nucleus could absorb several of them before it had a chance to decay. Hoyle's main contribution to the B^{2}FH collaboration (apart from providing the original inspiration for the work) was to calculate the r-process in all its glory, testing if the neutron captures and the subsequent radioactive decays did produce the observed abundances of the elements. In all of this, the team assumed that the flood of neutrons required to do the job came from the explosion of a massive star as a supernova at the end of its life. The approach was so successful that on the rare occasions when the abundance they calculated for a particular element turned out to be different from the one given in the published scientific literature for the actual abundance of that element, the published values proved to be wrong and had been reporting an inaccurate measurement of the actual abundance. In a sense, the success of this work "predicted" the existence of supernovae, since there was no other explanation for the source of the neutrons required for the r-process.

Although everybody on the team made contributions to all aspects of the work, and the surviving members are keen to emphasize this (especially in the wake of the Nobel fiasco), they did have their own areas of expertise. Fowler's work thoroughly reviewed the conversion of hydrogen to helium; Hoyle's theoretical work and the Burbidges' on the observational side (with Geoffrey Burbidge having a leg in both camps) explained the observed abundances of everything from carbon right through to uranium and beyond (information from nuclear bomb tests, declassified around the time this work was being done, showed that unstable radioactive elements even heavier than uranium had been formed in the nuclear explosions, a powerful indication of what the r-process could do). The only gaps in their scheme concerned the vast bulk of helium that is not made in stars, but had emerged from the Big Bang (a gap filled by the work of Hoyle and Tayler in 1964), and the origin of deuterium (heavy hydrogen), as well as lithium, beryllium, and boron (rare light elements with three, four, and five protons in their nuclei, respectively). It was the formation of these light elements in the Big Bang that was triumphantly explained by Wagoner, Fowler, and Hoyle in 1967, as I described in the previous chapter. With this solid grounding in nuclear physics and cosmology laid, we can now jump forward to the best modern understanding of how the physics operates inside stars to make the elements and scatter them through space.

8

The Superstar Connection

The Sun is not destined to play a major part in seeding the Galaxy with heavy elements. Even though it is a relatively massive star, in the sense that at least 90 percent of all stars are less massive than the Sun, it is still not massive enough to cook up anything heavier than carbon, oxygen, and a little nitrogen during its lifetime. To do more than that, a star has to start out with at least four times as much mass as our Sun — and to make all the heavy elements, a star has to start out with a minimum of about eight to ten times as much mass as our Sun. But still, carbon, nitrogen, and oxygen are extremely important elements, as we have seen, and we shouldn't ignore the lesser stars that make so much of the stuff. Even though the Sun will never manage to eject much of the material it does manufacture out into the Universe at large, stars very much like the Sun but in different locations do manage to manufacture even heavier elements, all the way up to iron, and eject a fair bit of this processed material into interstellar space.

The time a star spends on the main sequence depends on its mass — about 10 billion years for the Sun, 500 million years for a star with three times the Sun's mass, 200 billion years for a star with half the Sun's mass. During all that time, whatever its size, the star generates heat by converting hydrogen into helium in the way I have already described. When a star with about the same mass as the Sun has used up all the hydrogen in its core in this way, the core itself has to collapse, slowly, under its own weight. This does two things. First, it makes the core itself hotter, as gravitational energy is released. Second, hydrogen from outside the core falls inward into the core region, where it is hot enough for this "new" hydrogen to begin burning to make helium, by the CNO cycle, in a shell around the core. Both processes generate extra heat, and this heat flowing outward from the core makes the outer layers of the star expand, so that it becomes a red giant. The CNO cycle is particularly important because, as well as generating heat, it is the main source of nitrogen in the Universe. Although all three kinds of nuclei are involved, as the reaction progresses it shifts the balance of the chemical mixture in the affected region of the star away from carbon and oxygen and toward nitrogen.

But while this is going on in a shell around the core, the core itself continues to contract — as far as it can. There is a limit to how closely nuclei can be packed together, and when nuclei reach this state (which is determined by the laws of quantum physics), they are said to be "degenerate." To give you some idea of the densities involved, if the whole of the Sun, which has a mass 330,000 times that of the Earth, were in the form of degenerate nuclear matter, it would be almost exactly the same size as the Earth. The triple-alpha process begins to operate in the core of a star like the Sun only when the core has become a degenerate ball of helium nuclei, with a temperature of around 100 million degrees — for the Sun itself, this will happen about a quarter of a billion years after it stops burning hydrogen in its core and begins to convert itself into a red giant. But when helium burning does begin, for a star with the same mass as the Sun it happens in a flash, affecting the entire degenerate core.[1] The heat generated in the helium flash, which lasts for only a few minutes, expands the core away from the degenerate state and back to something like the deep interior of a star like the Sun, but with higher temperatures, densities, and pressures in the core. It also blasts a lot of material — perhaps 25 to 30 percent of the star's original mass — away into space from the outer regions of the star. For stars with at least twice as much mass as the Sun, helium burning switches on in a more gradual fashion, but the end result is much the same.

At the beginning of this stage of its life, a star like the Sun will have half its mass in the (no longer degenerate) helium core, and will be shining, overall, a hundred times brighter than the Sun does today. But its outer layers will have expanded into a huge gaseous atmosphere, inflated by the heat from below, so that the amount of energy crossing each square meter of the surface is quite low, and the surface is a cool red color. When the Sun reaches this stage of its life as a red giant, it will have expanded so much that its diameter will be bigger than the orbit of Mercury, but it will have lost at least a quarter of its original mass, as material from the outer layers escapes into space. This phase of a star's life doesn't last for long, though, because helium burning provides much less energy than hydrogen burning. The overall energy that is liberated when three alpha particles fuse to form a

1. The changes in the structure of the star at this point in its life literally take only a few seconds. When astronomers first developed computer programs to calculate how these changes would occur, in the 1960s and 1970s, it actually took much longer to run the programs than it does for stars to make the changes in their structure being described by the computer simulations (models).

carbon-12 nucleus is only 10 percent of the energy liberated when a helium-4 nucleus (an alpha particle) is made out of four protons (hydrogen nuclei), so helium would have to be burned much more rapidly than hydrogen just to maintain the brightness of the star, let alone keep it shining a hundred times brighter than that. Helium burning in a star with the same mass as the Sun actually lasts for only about 150 million years. While it is going on, the star will have two sources of energy — helium burning in the core and a thin shell of hydrogen burning still going on around the core.

Once again, the effect of all this activity on the outward appearance of the star is not what you might expect — this time, instead of expanding even more after the helium flash, the outer layers of the star shrink a little, with its luminosity dropping to about a tenth of what it was before the helium flash. The reason is that because the inner helium core of the star has expanded, the region available for hydrogen burning around the core is reduced, so this energy production process, although still active, is less effective than it was before the helium flash. But at least the logic is consistent — when the core shrinks, the outer layers expand, and when the core expands, the outer layers shrink.

Helium burning in the core doesn't only make carbon, because under these conditions carbon nuclei interact reasonably readily with alpha particles to make oxygen nuclei, which helps to delay the inevitable; so the "ash" produced by helium burning is a mixture of carbon and oxygen. But this is the end of the line for a star which starts out with about the same mass as the Sun. After all the helium in the core is used up in this way, it settles down as a cooling ball of degenerate material, because it never gets hot enough in its heart to trigger further phases of nuclear burning. During its later life, such a star will eject even more of its material from its tenuous outer atmosphere, as it is blown away into space; the cinder left behind by the Sun itself will be a white dwarf with only half of the Sun's original mass.

For many stars with masses between about one and four times the mass of our Sun, the outer layers are blown away almost entirely as a shell of ejected material around the central star, expanding away from it. These shells are visible as long as the central star continues to shine and light them up; they are called planetary nebulae, because when they were first studied through small telescopes, their roughly circular appearance made them look a little like planets.

Astronomers calculate that, on average, all the planetary nebulae in the Galaxy recycle about five solar masses of stellar material back into interstellar space each year. This is about 15 percent of all the material ejected by stars and available to be recycled to make new stars — most of this will simply be hydrogen and helium (especially for the smaller stars in the range), but in some cases heavier elements from the nuclear burning processes do get mixed into the outer layers of the star before they are expelled. This is one way in which oxygen, carbon, and nitrogen get into interstellar clouds after being manufactured inside stars. Indeed, the process of mass ejection from low-mass stars is particularly important as a source of nitrogen, which is a by-product of the CNO cycle; although some carbon and oxygen are formed in other ways (as we shall see), this is virtually the only source of nitrogen in the Universe. You can be absolutely sure that all the nitrogen in the air you breathe and in your DNA (along with most of the carbon in your body) had a previous existence as part of a planetary nebula, and was expelled from one or more red giant stars.

But after this phase of activity so crucial to the existence of life as we know it, the red giant has had its day, as long as it only started out with less than a few times as much mass as our Sun. The remaining core of the star, a degenerate ball of carbon and oxygen, will settle down quietly as a white dwarf and gradually fade into insignificance (indeed, stars with less than half the mass of the Sun never even get hot enough inside to trigger helium burning and end their lives as balls of degenerate helium stuff). At least, it will settle down as a white dwarf provided that the total mass of the remnant left behind after the outer layers have been expelled is less than 1.4 times the mass of the Sun. For anything more massive than this, a further stage of collapse is inevitable, as we shall see. But in round terms, allowing for the considerable mass of gas expelled during its time as a red giant, an old age as a cooling white dwarf is the fate of all isolated stars which start out with less than about four times the mass of the Sun.

A star with about four times as much mass as the Sun spends only a little over 600 million years on the main sequence and runs through the later stages of its life correspondingly quickly—as a (very rough) rule of thumb, the total time spent as a red giant (before and during helium burning) is about 10 percent of the time spent as a main-sequence star. But although their lives may be short compared with that of the Sun, stars with masses in the range from four to eight times the mass of the Sun end their lives in much more spectacular fashion, enriching the Galaxy in the process.

You might think that once all the helium in its core had been converted into carbon and oxygen, a more massive star than the Sun would simply shrink down upon itself once again, releasing gravitational energy and getting hotter inside until it became possible to generate energy by fusion of carbon and oxygen nuclei to make even heavier elements. But unless the star has at least eight to ten solar masses of material to hold everything in its core in place, life isn't quite that simple.

The first important factor is that even at the end of its life as a red giant, the remnant of such a star will (unlike the Sun) still have more than about 1.4 times the mass of the Sun left over in its former core, after all the hydrogen and helium in its atmosphere (and bits and pieces of things like carbon and nitrogen and oxygen) have been blown away. This matters, because there is a limit to the strength even of degenerate matter. If a white-dwarf stellar remnant has more than 1.4 times the solar mass (a number known as the Chandrasekhar limit, after the astrophysicist who first calculated the effect), then gravity overwhelms the quantum forces that make degenerate matter stiff, and the stellar remnant has to collapse rapidly in upon itself. This releases a great deal of heat in a hurry and triggers a furious bout of nuclear fusion.

In the final stages of the active life of such a star, nucleosynthesis is more complicated than the addition of alpha particles that took us from helium to oxygen. Carbon "burning" now involves carbon nuclei combining with one another in different ways, and ejecting various particles as they do so. Two carbon-12 nuclei can fuse to make one nucleus of neon-20, with an alpha particle left over; or they can fuse to make sodium-23, while ejecting a proton; or they can combine to make magnesium-23, while ejecting a neutron. This is where the neon in neon lighting, the sodium in common salt, and the magnesium used (appropriately) in fireworks comes from — carbon burning inside stars. As well as releasing energy (actually slightly more energy for each of these interactions than is released when a single carbon-12 nucleus is formed by the triple-alpha reaction), these interactions produce all the kinds of particles needed to interact with other nuclei (such as those of oxygen) to increase the variety of elements present. Under the extreme conditions that exist in the core of such a red giant (effectively a degenerate white dwarf), this produces a runaway series of nuclear interactions that proceeds all the way to nickel-56, which then decays to produce iron-56. For a star with four to six times the mass of the Sun, this may simply blow the outer layers away, leaving a "normal" white dwarf, below the Chandrasekhar limit, behind. But for a star with six to eight times the mass of the Sun (and possibly even for the slightly smaller stars), this explosive burst of nucleo-synthesis completely disrupts the star and scatters its entire mass out across interstellar space, in the form of heavy elements. The amount of iron spread across the Galaxy in a single explosion of this kind can amount to well over half the mass of our Sun, with about a quarter as much oxygen produced as iron, and lesser quantities of other elements. This scenario of complete disintegration of a degenerate star in a supernova triggered by explosive carbon burning was first suggested by Fred Hoyle and Willy Fowler in 1960, but has been considerably improved since, through the usual combination of improved theoretical models (based on better nuclear-capture cross sections and improved computer simulations) and observations of real supernovae. It turns out, as we shall see, that although an isolated star with about 8 solar masses of material can be disrupted in this way, the same kind of disruption occurs in smaller stars which are members of binary systems. But before going into details, I should try to convey some idea of just how big an explosion a supernova represents.

From the name, you might think that a supernova is like a nova, only bigger. That's true in a way — but only in the same way that a hydrogen bomb is like a firework, only bigger. Novae got their name because to the astronomers of ancient times they appeared to be "new" stars that suddenly flared into existence. In fact, we now know that novae are temporary outbursts of much fainter stars, which can often be seen using telescopes, and are not new at all. In a typical nova outburst, a star brightens by a factor of about one hundred thousand in a few days, then fades away back to its former level over a few months. In an ordinary galaxy like our Milky Way, there are about twenty-five novae each year. They occur in binary systems, where a white dwarf with a mass well below the Chandrasekhar limit is in orbit around a red giant. Material from the tenuous outer layers of the red giant is attracted by the gravity of the white dwarf and falls onto its surface, at a rate equivalent to about a billionth of the mass of the Sun each year. There, the hydrogen and helium mixture from the red giant builds up a layer on the surface of the white dwarf until the pressure at the bottom of the layer is great enough to cause an outburst of nuclear reactions, blowing the material away into space as the star flares up. The process can then repeat itself.

Impressive though the release of energy in a nova is by human standards, it is peanuts compared with a supernova, which releases a million times as much energy and briefly shines as brightly as all the stars in a galaxy like the Milky Way put together. It is literally, for a few weeks, as bright as a hundred billion suns. Supernovae are much more rare than novae — Tycho Brahe saw one in our Galaxy in 1572, and Johannes Kepler saw one just thirty-two years later, in 1604, but none has been seen in our Galaxy since then, although in 1987 a supernova was observed in the Large Magellanic Cloud, a small galaxy close to our Milky Way. Indeed, supernovae are so rare that astronomers began to appreciate their true nature only in the mid-1920s, with their initial realization of just how big the Universe is. Until then, it had still been possible to argue that the system we now call the Milky Way Galaxy, a flattened disk of stars about 90–100 thousand light years across and containing a few hundred million stars, was the entire Universe. Fuzzy patches of light in the sky, called nebulae, had been noticed long before this, but in the early twentieth century nobody could establish unambiguously whether these fuzzy blobs were clouds of material inside the Milky Way, relatively small star systems (like clusters) in orbit around the Milky Way, or (the most extreme possibility) entire galaxies of stars, like the Milky Way, but so remote from us that individual stars could not be picked out in them even with the most powerful telescopes available.

The discussion was confused by the discovery of what seemed to be an ordinary nova in one of these fuzzy nebulae (the Andromeda Nebula) in 1885. The "nova" was studied and photographed, but at the time there was no way of estimating its distance. Then, in 1901, another nova was seen in the Milky Way, and this time it was close enough for its distance to be measured, using a rather neat trick based on the speed with which light from the nebula lit up clouds of gas at different distances from the flaring star. Because we know how fast light travels, astronomers could tell how far these clouds were from the nova (obviously, if the light takes a week to reach a particular cloud, then that cloud is one light week away from the nova), and then use simple triangulation to work out how far the nova and the cloud must be from us. The technique is only approximate, because the clouds being lit up in this way are spread around the nova, with some a little closer to us than the nova, and some a little farther away, but approximate is better than nothing. It gave a distance estimate of 500 light years, almost next door in terms of the size of the Milky Way. But the "nova" in Andromeda had been 250 times fainter than the 1901 nova — which meant, if they were the same kind of object, that it must be about 8,000 light years away from us.[2] The Andromeda Nebula seemed to be a cloud of stuff within the Milky Way.

But when the technology improved and the 100-inch telescope on Mount Wilson became available, Edwin Hubble was actually able to measure the distance to the Andromeda Nebula in the mid-1920s, by

2. You have to multiply 500 by the square root of 250 to get the distance from the relative brightnesses, because brightness drops off as the square of distance.

identifying individual Cepheid variable stars in the nebula. He came up with a much higher value for the distance. With modern improvements in the technique, we now know that this "nebula" is indeed a galaxy very much like the Milky Way, rather more than two *million* light years away from us. This discovery was a key stage in developing a true understanding of the distance scale of the Universe, and eventually in working out its age (the time that has elapsed since the Big Bang), as I described in *The Birth of Time*. But what matters here is that if the Andromeda Nebula (or Andromeda Galaxy) is about 250 times farther away from us than was calculated in 1901, then the "nova" seen in it in 1885 must actually have been thousands of times brighter than the 1901 nova in the Milky Way and would have shone at least as brightly as 100 million suns.[3] The identification of this stellar outburst as something quite different from an ordinary nova was immediately confirmed when astronomers were also able to detect genuine novae in the Andromeda Galaxy and found that they really were as faint as the great distance implied. Indeed, it later transpired that these extra-bright novae are even brighter than the first calculation suggested, because the appearance of the 1885 event in the Andromeda Galaxy is dimmed by clouds of dust along the line of sight.

Remember that all this was still ten years before Hans Bethe (and others) worked out the nuclear fusion processes that keep stars shining. The extra-bright novae were referred to by a variety of names in the late 1920s, but the astrophysicist Fritz Zwicky began to use the term "super-novae," with a hyphen, in his lectures to students at Caltech in the early 1930s. After he and Walter Baade wrote a paper on

---

3. That a factor of 250 appears in this part of the discussion as well as in the earlier calculation is just a coincidence. All these figures are just round-number estimates to give you a feel for what is going on.

the subject in 1934, entitled *On Super-Novae,* the name stuck for good, and the hyphen was soon dropped. Reading that paper today, and in particular a second paper published by the same team in the same year, is almost as breathtaking as the first time you read B²FH. It isn't quite in that league, because the collaboration by Baade and Zwicky in 1934 was more speculative than definitive. But what else could it be? It was written years before astronomers had worked out in any detail what makes stars shine (ten years before Baade himself would trigger the interest of the young Fred Hoyle in how the heavy elements are built up inside stars), and only two years after the neutron had been identified by James Chadwick. It was, indeed, only three years after Subrahmanyan Chandrasekhar had published his calculations showing that any white dwarf star with more than 1.4 times as much mass as the Sun would have to collapse into some mysterious, and at that time unknown, form. Yet Baade and Zwicky leaped to the (correct) conclusion that the enormous output of energy from a supernova was associated with the collapse of an ordinary star into a superdense state, far denser even than a white dwarf, in which it was entirely made of neutrons — a neutron star. It was a brilliant example of scientific intuition, but also strictly in accordance with the logic spelled out by Conan Doyle through the voice of his creation, Sherlock Holmes: eliminate the impossible, and whatever you are left with, however improbable, must be the truth.

Baade and Zwicky published their ideas in two consecutive papers in 1934 in the *Proceedings of the National Academy of Sciences.* The first paper dealt with the problem of explaining where the energy released in a supernova explosion — which they estimated as tens of millions of times the rate at which radiation is released steadily by the Sun — could come from. The conclusion they reached from these calculations (after eliminating the impossible) was that "the total energy emitted in the super-nova process represents a considerable fraction of the star's mass." It was the second paper which pointed out that the most likely way in which energy comparable to the mc^2 of an entire star could be released was by gravitational collapse of the star into a very compact object. Gravitational collapse always releases energy — that, remember, is how stars get hot inside in the first place. But in order to release as much energy as they needed to power a supernova, Baade and Zwicky were forced to the conclusion that the end point of the collapse must be matter at its most extreme density, in the form of neutrons.

This wasn't quite a shot in the dark. In 1932, the Russian physicist Lev Landau, who was then on a visit to Niels Bohr's research institute in Copenhagen, seized upon the announcement that Chadwick had identified the neutron and immediately suggested to his colleagues there that stars might contain balls of neutron stuff in their hearts. He was thinking that the energy released steadily by a star over its lifetime might be produced by a gradual collapse of stuff from the outer layers of the star down onto this neutron core; but Baade and Zwicky were suggesting that a neutron star formed all at once, releasing all the available gravitational energy in a matter of a few days. "We advance the view," they said, "that a super-nova represents the transition of an ordinary star into a *neutron star*, consisting mainly of neutrons. Such a star may possess a very small radius and an extremely high density. [It] would therefore represent the most stable configuration of matter as such."[4]

Landau had returned to the Soviet Union, as it then was, in 1932, and didn't publish any of his ideas about neutron "cores" until 1938. About the same time, the American Robert Oppenheimer became

4. Their italics.

interested in the possible existence of neutron stars, and at the end of the 1930s, together with several of his students, he published a series of papers discussing the properties such objects must have, if the known laws of physics (especially the then-new ideas of quantum mechanics) were correct. The most dramatic conclusion of this work was that just as Chandrasekhar had shown that there was a limit to how much mass a white dwarf star could have without collapsing (presumably, it might have been thought by 1938, into a neutron star), Oppenheimer's team found that there is a limit to how much mass a neutron star can have without collapsing. But Baade and Zwicky were right — a neutron star does represent "the most stable configuration of matter as such." When an overweight neutron star collapses, it does so indefinitely — it disappears into what is now known as a black hole. The critical mass for a neutron star, now known as the Oppenheimer-Volkoff limit, is about three times the mass of the Sun.

None of these ideas were followed up immediately in the 1940s. One reason was that scientists became distracted from their academic work by World War II — Oppenheimer himself was a leading figure on the Manhattan Project, which led to the manufacture of the first nuclear bomb. But it was just as important for the development of astronomy (or rather, for the lack of development in this corner of astronomy at that time) that the theorists had raced ahead of the observers, coming up with ideas that simply could not be tested using the observational evidence available at the time. Although a few supernovae had been seen in pre-telescopic days, by people like Tycho, in 1934, at the time Baade and Zwicky first suggested that supernova explosions are powered by gravitational collapse, only twenty supernovae had been seen and photographed in modern times, and none had been studied in enough detail for their spectra to be analyzed. It was only in 1936, when a special kind of photographic telescope known as a Schmidt Camera became operational on Mount Palomar in California (where the new 200-inch telescope was being built), that Zwicky regularly began to find supernovae going off in galaxies beyond the Milky Way, at a rate of a couple a year. This was the start of the scientific study of supernovae; with the aid of the Schmidt Camera, which has a wide field of view, Zwicky could monitor many galaxies, and as soon as he spotted a supernova he could alert colleagues at nearby Mount Wilson, whose telescope (still the best in the world) could take their spectra.

Again, just when things were getting interesting, the project was curtailed by World War II. But at the end of the 1940s a larger Schmidt Camera became operational, as did the 200-inch telescope, and many more supernovae were discovered in the years and decades that followed. By the time Zwicky died in 1974, more than four hundred of these objects had been observed in some detail, Zwicky having discovered more than a quarter of them. With hundreds of supernovae having been photographed and analyzed spectroscopically, in the late 1970s and into the 1980s astronomers at last began to get a clear picture of what was going on, and of the fact that there is more than one kind of supernova.

The key distinction is between two kinds of supernova, known as Type I and Type II (there are subdivisions, of interest to the specialists, which I won't bother with here). The important thing for the story I have to tell here is that the two kinds of supernova produce different quantities of the various elements, and that both have contributed to the mixture of stuff that we are made from. The story emerged only slowly, from the usual combination of improving observations, based on better telescope technology, and improving theoretical models, based on computer simulations of what goes on in-

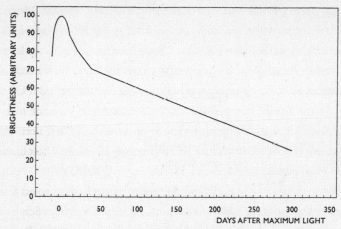

A schematic light curve showing the typical behavior of a Type I supernova.

side these stars at the end of their tether. But it was more or less complete by the mid-1980s.

The distinction between the two types of supernova is based on observed differences in their behavior. The way in which any variable star brightens and then dims is called its light curve, and the light curves of all Type I supernovae look much the same (if any of those specialists happen to be reading this, from now on when I refer to Type I supernovae, strictly speaking I am talking only about Type Ia supernovae). They show a rapid rise to maximum brightness, taking about two weeks, immediately followed by a steady decline over the next couple of weeks, tailing off into a more gradual exponential dimming of the brightness. In this exponential dimming, it takes a certain number of days for the star to fade to half its starting brightness, then the same number of days to fade to half of that brightness (one quarter of the starting brightness), and so on. The half-life of a Type I supernova's light curve is about fifty days.

As well as this distinctive overall behavior of their light, the spectra of Type I supernovae are like no other kind of star. When they are at their brightest, they show no sharp lines corresponding to the presence of atoms of particular elements but, rather, very broad bands of light and dark. This is interpreted as showing that the light is coming from a mixture of material moving in a violent and turbulent way, with individual atoms moving rapidly at random, so that the light from any individual atom may be enormously blueshifted (if it happens to be heading our way) or enormously redshifted (if it happens to be traveling away from us). So the normal pattern of light and dark lines in the spectrum gets smeared out by huge Doppler shifts into broad bands of light and dark. The random speeds required to do this are at least 10,000 kilometers per second—about ten thousand times faster than the random speeds of the molecules of the air you breathe. Intelligible spectra only begin to emerge from the resulting blur of light and dark bands as the material cools and the supernova fades from its peak brightness.

The light curves of Type II supernovae are not all exactly the same, but the key thing is that none of them looks like a Type I supernova. Overall, Type II supernovae don't just brighten and then immediately start to dim, but stay at around the maximum brightness for some time—perhaps weeks—and then fade away more slowly than Type I supernovae. Their spectra are also different. Although the atoms involved are moving fast enough to broaden the lines considerably, they don't move as fast as the atoms in Type I supernovae (in round terms, about one tenth as fast), so the spectral lines can still be identified relatively easily, even at maximum brightness. There is always a lot of hydrogen in the material being blasted out from a Type II supernova, as you might expect; but there is also a lot of other stuff, including helium, magnesium, and silicon. Significantly, when the spectra of

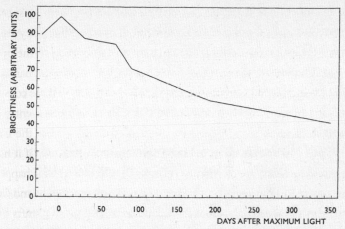

A schematic light curve showing the typical behavior of a Type II supernova. The key feature here is that the decline in brightness is more gradual than that of a Type I supernova; other differences between the two kinds of stellar explosion are discussed in the text.

Type I supernova can be unraveled, there is no evidence for hydrogen associated with the explosion, even though hydrogen is by far the most common element in the Universe. But we do see evidence of other elements as the Type I supernova cools down, including, crucially, iron.

And there is another difference between Type I and Type II supernovae. Type I supernovae occur anywhere in a spiral galaxy like the Milky Way, and also in the galaxies known as ellipticals. But Type II supernovae are seen only in the arms of spiral galaxies, among the clouds of dust and gas that are known to be associated with the birth of new stars. Finally, Type I supernovae are much brighter than Type II supernovae — all Type I supernovae have the same brightness, and this is typically three to ten times brighter than a Type II supernova, because the Type IIs don't all have the same brightness. This makes

Type I supernovae easier to spot in faint galaxies far across the Universe. Curiously, though, Type II supernovae emit much more energy than Type I supernovae do. It's just that most of the energy released in a Type I supernova shows up as visible light, while most of the energy released in a Type II supernova emerges in other ways. It is time to explain just why these two kinds of stardeath are so different from one another.

Type I supernovae occur in binary systems very similar to the ones in which ordinary novae occur. The crucial difference is that in the precursors to Type I supernovae, material is falling onto a white dwarf star which is already almost at the Chandrasekhar limit of mass — 1.4 times the mass of the Sun. In other circumstances, it would take a very long time for an isolated star with only about the Chandrasekhar mass to evolve to the point of being a white dwarf. But because these stars are in binary systems, stellar evolution can effectively be speeded up — the way each star ages is affected by the presence of a nearby companion star. The bigger star in the binary system will run through its life cycle more quickly than the smaller star and become a red giant. But when it does so, the gravitational pull of the companion, which has not yet reached that stage in its life cycle, will strip mass away from the swollen outer layers of the giant star, leaving the core, rich in the carbon and oxygen produced by helium burning, behind. The system contains a white dwarf, even though it is not very old, by stellar standards. Now, the star that was the smaller companion may have become more massive than this white dwarf remnant and will become a giant in its turn, dumping material back onto the carbon-oxygen white dwarf. Astronomers can actually see this process at work in some binary systems, because the infalling material gets so hot that it emits X-rays, which we detect, where it strikes the white dwarf. The white dwarf gains in mass as a result, and just at the point where it reaches the Chandrasekhar limit, the core of the star starts to collapse, but immediately becomes so hot that the carbon starts to fuse, releasing more energy and triggering a wave of nuclear fusion which races through the entire star like a high-speed flame, disrupting it completely and blowing all the material of the star — all the material produced by this burst of fusion — out into space, very much in the way Hoyle and Fowler suggested back in 1960.

But the modern understanding of Type I supernovae goes much further than those early speculations by Hoyle and Fowler. It is because this process always occurs in the same way, at exactly the Chandrasekhar mass, that all Type I supernovae look the same and reach the same maximum brightness. But since it may take hundreds of millions of years to build up the mass of the white dwarf from below the Chandrasekhar limit to the critical value, and since it may have taken even longer for the binary system to evolve to the point where mass is being transferred from a giant star onto a carbon-oxygen white dwarf, Type I supernovae are typically associated with old stars that may occur anywhere in the Galaxy. The process of semiexplosive nuclear burning in such an event is so intense that it goes all the way from carbon and oxygen to iron-group elements in the short space of time that the nuclear "flame" is racing through the material of the white dwarf. The nuclear reactions proceed all the way to nickel-56, which decays into cobalt-56, which in turn decays into stable iron-56. It is the radioactive decay processes, following an exponential law, which continue to release energy in the supernova remnant after the initial peak intensity has passed and which produce the characteristic half-life of the declining light curve. Overall, the energy released in a Type I supernova is very close to the amount of nuclear energy that would be released when about two-thirds of a solar mass of carbon and oxygen is converted into iron. About half of the mass of the original white dwarf is turned into iron in this way, with smaller quantities of other elements such as silicon and sulfur also scattered into space in the blast. Crucially, though, a Type I supernova does not manufacture any of the elements heavier than iron. Which brings us to the story of Type II supernovae.

Type II events occur, as I have mentioned, in stars which start their lives with more than about eight to ten times the mass of the Sun. These are the stars which live fast and die young—remember that a star with four times the Sun's mass has a main-sequence lifetime of about 500 million years, while a star with 20 times the mass of the Sun will stay on the main sequence for only a few million years. This is why we see Type II supernovae only in the dusty disks of galaxies like the Milky Way, in the regions where star formation is still going on. Stars that live only a few million years do not have time to move far from their birthplaces before they die—to put this in perspective, it takes the Sun about 250 million years to orbit around the Milky Way once, a journey it has already completed about eighteen times since it was born. But a 20-solar-mass star would not have time to complete even 1 percent of a single circuit of the Milky Way before it exploded.

On the way to that explosive death, such a star will run through the entire range of possible energy-liberating fusion reactions, from hydrogen up to iron-group elements. It does so in stages, in the way I have already outlined, with each fuel being fused in the core of the star in its turn, to be followed by a collapse which takes the temperature up to the point where the next phase of nuclear burning can begin. But each time, just as in the case of a shell of hydrogen-burning material surrounding the helium-burning core of a low-mass red giant, there will still be conditions farther out from the center of the star where the earlier phases of nuclear burning can go on. By the time a massive star has evolved to the point where a core of iron is building

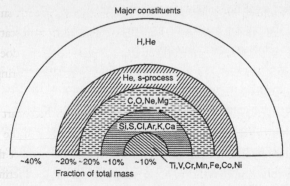

Major constituents

H,He

He, s-process

C,O,Ne,Mg

Si,S,Cl,Ar,K,Ca

Ti,V,Cr,Mn,Fe,Co,Ni

~40% ~20% ~20% ~10% ~10%
Fraction of total mass

The "onion shell" structure of the deep interior of a massive star just before it becomes a supernova.

up in its heart, it is surrounded by a series of shells in which other nuclear reactions are going on, wrapped tightly around the core like onion skins. Just outside the iron core, silicon is being converted into iron; in the next shell, oxygen (and some neon) is "burning" to make silicon; a little farther out, carbon is being converted into oxygen; in the next layer the triple-alpha process is converting helium into carbon; and on top of all that activity is a layer where hydrogen is still being converted into helium.

For a star with between fifteen and twenty times as much mass as the Sun, near the end of its life the iron core may contain rather more mass than our Sun, but with roughly the diameter of the Earth, like a white dwarf with the thin layers of fusion activity wrapped relatively tightly around it. These onion skins are themselves mixtures of stuff — the silicon layer also contains traces of sulfur, argon, chlorine, potassium, and calcium, while the oxygen layer also contains traces of neon and magnesium. The whole star will be at least fifty times as big as our Sun is now — perhaps even bigger, depending on how much (or how little) of its original atmosphere has been lost. Because of this mass loss, it will no longer be as massive overall as it was, with perhaps two or three solar masses of material (probably rich in nitrogen) having been blown away into space.

But the final stages of nuclear fusion do not last long. For anyone used to the kind of timescales involved in astronomy, the later stages of nuclear burning in a massive star occur with breathtaking speed. For a star which starts out with about seventeen to eighteen times as much mass as our Sun, after a few million years on the main sequence, helium burning will have kept the red giant shining for about a million years, carbon burning would do the job for only twelve thousand years, the energy released by neon and oxygen between them could have held the outer layers up for about ten years, and the silicon would have been burnt out in a few days. Then, things start to get interesting.

Because it takes energy to manufacture nuclei of elements heavier than iron by fusing lighter nuclei together, once the core of the star is converted into iron it can no longer draw on nuclear fusion to provide the energy to hold the star up. Indeed, it can no longer hold itself up. For millions of years, fusion has held gravity at bay. Now, gravity gets its revenge. The iron core collapses dramatically, in less than a tenth of a second. The collapse releases gravitational energy, but this doesn't go to heat the star. Instead, it is turned into kinetic energy which is used to smash apart the iron nuclei in collisions, undoing all the work of nuclear fusion, converting them back into a mixture of protons and neutrons. Since this absorbs energy (roughly speaking, it absorbs as much energy in a fraction of a second as the star has radiated during its entire previous life), it makes the core of the star cool, and this encourages the collapse (which releases even more gravitational energy). The process is so violent that it takes the core well past the white dwarf stage, and under the conditions of extreme density and

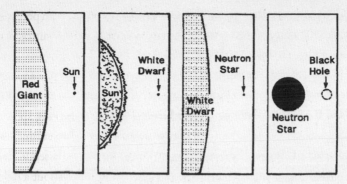

The relative sizes of stars.

pressure that are produced, electrons are forced to combine with protons to make neutrons — the reverse of the usual process of beta decay, in which a neutron spits out an electron and becomes a proton. For each neutron formed in this way, a single neutrino is released. The number of neutrinos released, in a fraction of a second, is equal to the number of protons in the iron core at the start of the collapse — about 10^{57}, a number so big that it is hard to put it in any kind of perspective. But try this. The energy released in a Type II supernova is about a hundred times as much as the entire energy output of the Sun over its entire lifetime. But only 1 percent of this energy emerges in the form of visible light. The other 99 percent is carried by the neutrinos. And *all* this energy comes from the gravitational energy released when a ball of stuff with about as much mass as the Sun collapses from the size of the Earth to the size of the island of Manhattan.

If the collapsing ball of stuff has more than three times as much mass as our Sun at this stage, nothing can stop the collapse. It becomes a black hole, the ultimate triumph of gravity over matter. But in the vast majority of massive stars, including most of the ones that become Type II supernovae, this is not the case. Instead, the collapsing ball of neutrons is abruptly brought to a halt, as quantum processes stiffen the material and prevent the neutrons from merging into one another in an amorphous blob. Indeed, the stiffening happens so suddenly that the freshly made neutron star — for that is what it now is — bounces back out a little, like a golf ball that has been squeezed in an iron hand and then released suddenly, before settling down into a stable state, with an amount of matter comparable to the mass of our Sun packed into a sphere less than ten kilometers across.

The latest computer simulations suggest that this is a two-stage process. The whole core collapses suddenly (in a few tenths of a second) to a ball of nuclear stuff about one hundred kilometers across. At this point, a bit less than half of the material, in the very center, has reached such extreme densities that the quantum stiffening process occurs and makes the inner core bounce. The bounce sends a ripple out through the material just outside the innermost neutron core, which has been falling in with it. "Ripple" is hardly the word to do justice to the shock wave produced by the recoil of this infalling material, traveling at about 15 percent of the speed of light, from the vibrating neutron core. As the neutron star settles down, taking several seconds for the whole neutron ball to shrink down to a diameter of about ten kilometers, almost all of the inward momentum of its collapse is converted by the bounce into outward momentum in the shock wave now racing *outward* from the core of the star.

But while all this is going on, the outer layers of the star — the fifty or more solar radii of more tenuous material, weighing a dozen times more than the Sun — have scarcely noticed what is going on. The complete core collapse takes, roughly speaking, a few seconds. But it would take several minutes for the inner layers of the outer part of the star to fall into the hole that has appeared beneath them, and while the neutron star is forming, they are essentially hanging unsupported above the void. As the theorists who study supernovae are fond of pointing out, the situation is like that of a cartoon character who runs off a cliff and hangs motionless in the air until he notices what has happened. In the case of a Type II supernova, just as the outer layers of the star begin to fall, they are hit from below by the outward moving shock wave, which tries to push them upward and out of the way.

On its own, the shock wave would never succeed in doing this. As it plows into the upper part of the star, it piles up material in front of itself, like a snowplow trying to force its way through a mountain pass that is completely blocked by a huge drift of snow. The shock wave, trying to push a dozen solar masses of star stuff out of the way, slows as the density of the material in the shock builds up, and it would come to a halt very soon except for one thing — or rather, a huge number of very small things. The second stage of the collapse, taking the neutron core from a diameter of about one hundred kilometers down to a diameter of about ten kilometers, involves an enormous release of gravitational energy, which is converted into heat and raises the temperature of the neutron star to about a hundred billion degrees. Under these conditions, heat energy appears in the form of gamma rays, not visible light, and the gamma rays are converted into electrons and positrons (in line with $E = mc^2$). Many of these particles are then involved in interactions which produce neutrinos — a few times more neutrinos, it turns out, than those 10^{57} produced when all the protons in the core were converted into neutrons. In the ten seconds or so that it takes for the neutron core to complete its collapse, so many neutrinos are produced that they carry away more than a hundred times as much energy as the energy involved in the material explosion of the star. They all stream outward at very nearly the speed of light, most of them passing through all the outer layers of the star and away into space.

The important point, though, is that not all of them do. Neutrinos are famously reluctant to interact with anything, and although there are about a billion of them in every cubic meter of the Universe (including every cubic meter of the room you are sitting in), we just don't notice them in everyday life. If a beam of neutrinos were to travel through a wall of solid lead 30,000 light years thick, only half of them would be absorbed by nuclei of lead along the way. Yet the shock wave trying to move out through the dying star would be so dense that it would absorb a significant number of the neutrinos from the core, giving it a boost which sends it moving on at about 2 percent of the speed of light, eventually blowing the entire outer layer of the star (at least half of the star's original mass) away into space. Along the way, a flood of nuclear reactions involving neutrons would have taken place in the shock itself, manufacturing very heavy elements from the r-process. The rest of the neutrinos, traveling at very nearly the speed of light, would have zipped through the remainder of the atmosphere of the star and escaped into space before anyone watching the star from outside would have noticed anything unusual going on. The neutrinos reach the surface of the star about two hours before the shock wave, moving at "only" one-fiftieth of the speed of light, reaches the surface. And it is only when the shock wave reaches the surface of the star that it becomes visible as a supernova. But it's the elements themselves we are interested in here, rather than the details of how a supernova works. At last we have found the place where things like copper, uranium, silver, mercury, and lead are manufactured. But don't think that they are produced in anything like the quantities of lighter elements made during the earlier stages of the life of a star. Remember that hydrogen and helium together make up 99 percent of all the mass in the Universe that exists in the form of atomic nuclei. All the elements from lithium (with three protons per nucleus) to the iron group (with twenty-six protons per nucleus) together make up less than 1 percent as much mass as hydrogen and helium put together. Yet those elements are common compared with everything else — all the mass of all the nuclei in the Universe with more than twenty-six protons in each nucleus amounts to less than one-thousandth of the mass of everything from lithium to iron. And if you leave nickel-28 out of the calculation, the heavy elements contribute only one ten-thousandth of the mass of all the light elements except hydrogen and helium.

There's one other quirk in the way elements are manufactured and distributed by supernovae that is worth mentioning. We already saw that Type I supernovae are very good at spreading iron into space. But Type II supernovae, even though they are triggered by the collapse of an iron core, hardly release any iron into space at all — it has all gone into making the new neutron star. On the other hand, the outer layers of the progenitor of a Type II supernova are rich in oxygen, and this gets blown away into space by the shock wave. The kind of Type II supernova I have just described will liberate about 1.6 solar masses of oxygen, while, as I mentioned earlier, a Type I supernova liberates about two-thirds of a solar mass of iron. The actual amount of iron and oxygen that we see around us, in the Universe at large but in particular in the Sun itself, is a mixture resulting from both kinds of process, and the fact that this mixture is not dominated by either iron or oxygen confirms that both kinds of supernova were at work in the Milky Way long ago when the Sun formed.

This is an important point. Even though supernovae are relatively rare events, and have not been studied in the same detail as ordinary main-sequence stars, the observational evidence and the theoretical models do indeed match up to one another. The most impressive confirmation of this came in 1987, when a supernova was seen to explode in the Large Magellanic Cloud, one of the star systems that is a companion to our Milky Way Galaxy. This was the nearest supernova to have been observed since the invention of the astronomical telescope, and every available astronomical instrument was turned on it to study the event and its aftermath.

The Type II event, known as SN 1987A because it was the first supernova observed in 1987, involved the explosion of a star with about seventeen to eighteen solar masses (which is why I picked that size in the example above), about 160,000 light years away from us (so by the time we saw the supernova explode, the actual star involved had been dead for 160,000 years). The behavior of the supernova closely matched the predictions based on computer models and observations of more remote supernovae over the years — not perfectly, so there is still more work to do on those models, but closely enough to be sure that the picture I have just painted for you is accurate in broad outline. The observers were able to identify the star that had exploded using old photographic survey plates taken before it blew up, so they could identify just what kind of star it was (which is how we know its mass). And, in a huge bonus, quite by chance there were several experiments running on Earth at the time the supernova was seen that could detect neutrinos from space. When the data from these experiments were analyzed (after the supernova had been seen), they showed that a handful of neutrinos from the blast had been stopped in the detectors on Earth, less than three hours before the supernova had been seen to explode. This was dramatic confirmation that neutrinos from the core collapse had sped out through the outer layers of the star, giving the shock a boost along the way, and had escaped into space before the shock wave reached the outer regions of the star, and those neutrinos provided a direct window on the events going on during the core collapse.

They also give us another way to try to put the vast number of neutrinos released in the core collapse in perspective. Altogether, from the conversion of protons into neutrons and the other processes going on in the collapsing core, SN 1987A produced about 10^{58} neutrinos. Imagine them spreading out in all directions through space, as an expanding spherical shell around the supernova. At the distance of the Earth, they would fill a thin shell about 10 light seconds thick and with a radius of 160,000 light years. Even thinned out in this way, at the distance of the Earth from the supernova this still left 100 billion neutrinos to pass through every square centimeter of the surface of the Earth (and through every square centimeter of you) in the space of about ten seconds. Neutrinos are so reluctant to interact with ordinary matter that out of all this flood of particles, on average one person in every thousand, out of the entire population of the Earth, had a single neutrino from SN 1987A stop in his or her body. Indeed, only twenty-two neutrinos from the supernova were stopped in the detectors specifically built to trap such entities — but this is no less than is required to match our understanding of how supernovae (and neutrinos) work. In a triumphant confirmation of the accuracy of the models of how the neutron star formed, that pulse of neutrinos arrived spread over a span of just twelve seconds, very close to the calculated duration of the core collapse.

Even this, though, wasn't the best piece of observational evidence to come from SN 1987A, in the eyes of the theorists who had spent decades studying the way the chemical elements are manufactured inside stars and spread throughout the Milky Way. The standard theory of supernovae, developed in the years before SN 1987A was seen to explode, said that almost all the energy radiated by the supernova as visible light during the first hundred days of its life would come from the decay of cobalt-56, produced in the early stages following the blast, to iron-56. This is the second step of a two-step process, remember, since the iron-group element produced in most profusion directly in the blast itself is nickel-56, which decays, on the familiar exponential timescale, with a half-life of just over six days to make the cobalt-56. The cobalt-56 decay, which has a half-life of seventy-eight days, then dominates the energy production in the fading supernova remnant over the next few months. The detailed shape of the light curve of SN 1987A showed that in the first hundred days of its life after reaching peak brightness, 93 percent of the energy was indeed being produced by the decay of cobalt-56.

How much cobalt was involved in this decay? Analysis of the fading light curve of supernova 1987A showed that, in total, it produced a mass of cobalt-56 equivalent to about 7 percent of the mass of our Sun, or twenty-three thousand times the mass of the Earth — which sounds impressive until you recall that the total mass of the star at the end of its life was about fifteen times the mass of the Sun. So the proportion of its mass that was transformed into radioactive cobalt-56, and kept the supernova shining brightly while it was studied by astronomers on Earth, was only about half of 1 percent of the mass of the star itself. Once again, this closely matched the predictions made by the theorists.

Observations of the fading supernova continued well into the 1990s — indeed, they are still continuing — and this exponential decay continued smoothly until January 1990, five hundred days after the supernova was first observed, when the star suddenly started fading more quickly. There must still have been plenty of cobalt-56 around at that time, still decaying to iron-56, and the explanation is that this dimming of the light from the supernova marked the time when tiny solid particles began to condense out of the maelstrom of material expanding away from the site of the explosion, forming a kind of soot of tiny grains which blocked some of the light. Again, this is exactly what the models predict. Finally (for now) about one thousand days after the supernova was seen to explode, the light curve began to flatten out into a more gradual decline. This marked the time when most of the cobalt-56 had decayed to stable iron-56, and the continuing energy source in the supernova remnant became rarer, but longer-lived, radioactive nuclei such as cobalt-57 (with a half-life of 271 days) and titanium-44 (with a half-life of some 47 years). To put this source of energy in perspective, though, seven hundred days after it was seen to explode as a supernova, the stellar remnant was glowing more faintly than it had in its last days as a giant, just before the explosion.

While these overall observations of the declining light curve were being made, astronomers were also getting to witness light from successive layers of the star, being expelled into space one after another in a kind of cosmic striptease, and were able to study them spectroscopically. These studies showed the presence of nickel-56 directly, in the early days after the supernova was seen to explode, and confirmed that a mass of nickel-56 equal to 8 percent of the mass of our Sun had been manufactured in the supernova, again closely matching theoretical predictions. The spectroscopic studies also revealed the presence of barium, strontium, and scandium — all s-process elements produced before the star became a supernova — now being expelled into space. The agreement between all these observations and the predictions of the theoretical models was so impressive that it leaves no room to doubt that the elements really are manufactured inside stars and spread throughout the Milky Way in supernova explosions, and that we do understand not just the broad outlines of the process but considerable detail about what is going on. Don't just take my word for it. Roger Tayler maintained his interest in the origin of the elements after his work with Hoyle at the beginning of the 1960s and became an authority on how the chemical composition of a galaxy like the Milky Way is altered as the elements are cooked inside stars and distributed through space. Later in the 1960s, he moved to the University of Sussex, where he spent the rest of his career, and where I learned pretty much everything I know about nucleosynthesis and the evolution of the chemical elements from him. In the early 1990s, he told me that in his opinion the spectroscopic observations of the material expelled by SN 1987A, coupled with the detailed shape of the decline of the light curve, were "the most important and exciting ones [he had ever seen] concerned with the origin of the elements, confirming that the theoretical model is broadly correct."

We know where the elements came from and why they exist in the Universe in the proportions they do. But how do they get from the supernovae into stars like the Sun, planets like the Earth, and people like us? The clue lies in the way that SN 1987A faded abruptly at the beginning of 1990, as it became shrouded in a cocoon of solid grains of material — stardust.

9

Sowing the Seed

Stardust is the key to the existence of complex molecules in the Universe, and therefore to the existence of life itself. Tiny grains of solid material expelled from stars — either gradually, as a red giant sheds its outer layers, or violently, in nova or supernova explosions — provide both the sites where interstellar chemistry can occur and the seeds which carry the resulting complex molecules from one part of the Galaxy to another. But none of this was appreciated until astronomers developed their understanding of how the elements formed. It was only in the late 1960s, little more than thirty years ago, that complex molecules were first identified in space, by their characteristic spectroscopic signatures at radio wavelengths. The first identifications, of ammonia and water vapor, were interesting, but not really surprising (and barely counted as polyatomic molecules, with only four atoms in each molecule of ammonia, NH_3, and three in each molecule of water, H_2O). But in 1969 the organic molecule formaldehyde was identified in this way. "Organic" means that the molecule contains carbon (its chemical formula is H_2CO), and formaldehyde is also a molecule that is associated with the emergence of life on Earth, a chemical building block which commonly appears as a subunit in more complex organic molecules, including sugars involved in the life processes going on in your body. The presence of formaldehyde among the cold clouds of gas and dust in space hinted at the possibility of a rich complexity of interstellar chemistry, and subsequent discoveries have fulfilled that promise, identifying well over a hundred polyatomic molecules in space, many of them containing many more atoms than a molecule of formaldehyde. They include long chains in which up to eleven carbon atoms are linked in a row, with a hydrogen atom at one end and a nitrogen atom at the other, rings known as polycyclic aromatic hydrocarbons (PAH), and such relatively familiar compounds as ethyl alcohol, formic acid, and hydrogen cyanide. The PAH molecules (also sometimes referred to as polyaromatic hydrocarbons) are especially interesting, because they are the most stable hydrocarbon molecules under the conditions that exist in these interstellar clouds. They are so large that they are called macromolecules, and each is made up of several rings of six carbon atoms, forming a structure of little hexagons joined together along their edges, with perhaps a hundred or so carbon atoms altogether, and hydrogen atoms attached round the free edges of the outside rings. One or two additional varieties of polyatomic interstellar molecules are discovered every year. But how do they form?

Some of the simpler molecules are relatively easy to make out of mixtures of gases — hydrogen and oxygen, for example, react enthusiastically to make water. But others need a surface to stick to, something like a tiny grain of carbon (in the form of graphite), which sweeps up atoms from the cloud as it moves through the gas. Atoms stick to the surface of a grain, where they easily interact with one another. If a polyatomic molecule did start to form in the gas itself, the impact of another fast-moving atom would be more likely to break it apart than anything else. But on the surface of a dust grain the existing molecules are not likely to be broken apart by the impact of another atom, because the grain itself absorbs the impact and gives the incoming atom a chance to stick to the growing molecule.

Astronomers have known for a long time that there are large amounts of dust in many of the cold clouds of material in space, and the observations of the declining light curve of SN 1987A provided a nice confirmation of this. But at first sight the presence of so much dust around a young Type II supernova was surprising.

The puzzle is that under the conditions that exist in space, an energetic mixture of carbon and oxygen is extremely eager to react, the two elements combining to form the gas carbon monoxide (CO). If such a mixture of carbon and oxygen is ejected by a star, then the element which is least abundant ought to be used up in this reaction, leaving the remainder of the other one left over to take part in other chemical reactions. In stars which expel a lot of carbon and only a smaller amount of oxygen, you would naturally expect to see carbon dust as a result. But in Type II supernovae, like SN 1987A, a lot more oxygen is produced than carbon — so why isn't all the carbon locked up in carbon monoxide molecules?

The answer seems to be that any carbon monoxide molecules that formed in the shell of material expanding away from the supernova were ripped apart by highly energetic electrons (beta rays) produced in the radioactive decay of cobalt-56 that kept the supernova shining so brightly for so long. This gave the carbon atoms a chance to condense to form grains of graphite dust, even though there was more oxygen than carbon around in the supernova debris. But the dust is very fine indeed — the particles of interstellar dust are individually about the same size as the solid particles in a cloud of cigarette smoke.

There is no doubt that these tiny pieces of supernova debris are formed in stars and spread across the Galaxy. They have even turned up here, as little pieces of grit trapped in meteorites that have fallen to the surface of the Earth. The bits of grit really are tiny — typically, a few millionths of a meter (that is, a few thousandths of a millimeter) across. But they can be sliced open and their composition analyzed, which in many cases shows that they contain just the proportions of isotopes that the theoretical models predicted for material cooked inside stars. For example, a high proportion of the isotope carbon-12, relative to carbon-13, combined with a lacing of silicon-28, is a clear sign that a graphite particle has formed in the immediate aftermath of a supernova explosion. Although the particles are too small to see with the naked eye, they are visible under the microscope and could, in principle, be touched — it is possible to hold a piece of pure supernova debris in your hand, even if you wouldn't feel it nestling there.

You could even go one better than that, if you knew the right people. Some of the tiny carbon grains found in samples of material from space aren't in the form of graphite, but diamond. Diamond crystals are a form of carbon which is produced under intense pressure — there is actually some scientific basis for the way Superman, in the stories, converts an ordinary lump of coal into a diamond simply by squeezing it hard with his superstrong hand. Diamond crystals from space are the result of intense squeezing of graphite grains occurring in regions of an exploding supernova shell where the pressure has reached rare peaks of intensity as shock waves have passed through it. It is a sign of just how sensitive and accurate the measuring equipment used by physicists these days has become that this explanation for the origin of diamonds is confirmed by analyzing traces of the element xenon found in them—even though this material is so rare that only one diamond grain in a million actually contains even a single atom of xenon. The mixture of isotopes of xenon found in diamonds could not be produced by any single nuclear process—but it is exactly the mixture expected from a combination of the products of the p-process and the r-process at work. Since theory tells us that these two processes operate at different levels in an exploding supernova, the discovery also tells us that the stuff from the supernova gets thoroughly mixed up in the explosion. Overall, the study of diamonds from space confirms how well astrophysicists understand supernova explosions, as well as confirming that grains of stardust (in this case, grains that would literally twinkle like little stars) can cross space and end up in the material from which new stars and planetary systems form. The interstellar medium certainly is enriched with material processed inside stars. But just how much of this material is around in a galaxy like our Milky Way?

"Empty space" really isn't empty, even though it corresponds to a vacuum far more devoid of atoms and molecules than the kind of "vacuum" physicists work with in laboratories here on Earth. On average, there is about one hydrogen atom in every cubic centimeter of the space between the stars of the Milky Way. In places we can see dark clouds of dust, obscuring the light from the stars beyond them, so that they look like dark tunnels through the Milky Way. Because these dark clouds are cold (no warmer than about 10–15 K, roughly 260 degrees below zero on the Celsius scale) they do not radiate much energy. But they have a key role to play in the story of interstellar chemistry.

Although these clouds are so cold today, the grains in them formed from very hot material in the aftermath of a supernova explosion (or probably from several supernova explosions, with the resulting debris now thoroughly mixed up). Because oxygen is the most common element around after hydrogen and helium, it is very easy for oxides to form in the original mix of material, and these solidify in turn — changing straight from the gas state to the solid state — as the gas cools. The behavior of oxides under these conditions is well known from studies in laboratories here on Earth, and we know that aluminum oxide particles will be the first to condense out in this way, followed in turn by the oxides of calcium, titanium, nickel, iron, magnesium, and silicon. Although silicon oxides are not the first to condense out, silicon plays a particularly important role in what follows, both because it is a very common element by the standards of interstellar space (although not as common as CHON) and because silicon oxides can combine with the oxides of other material present to form silicate grains. A silicate molecule contains a group of one silicon atom and four oxygen atoms linked to form a silicate group (SiO_4) which is linked to something else (such as aluminum or magnesium) in a silicate, but which acts as a single unit in many chemical reactions. Silicates are common in space for the same reason that oxides are — there is an ample supply of silicon oxides around to combine with just about all the other oxides (except carbon monoxide) and lock them up in silicates. One sign of this is that silicates make up about 90 percent of the material of the rocks of the Earth's crust, another link between ourselves and our cosmic origins.

Along with graphite grains, silicates are particularly important in the next stage of cooling in the aftermath of a supernova explosion, when icy shells form around the solid grains. "Icy" includes all kinds of ice, not just frozen water but also frozen methane, frozen ammonia, and even frozen carbon monoxide. It is this frozen mixture of ices, wrapped around a core of graphite or silicate material, that behaves like a tiny cold test tube in which the chemical reactions that make the variety of polyatomic molecules detected in space take place. Although the icy particles are by now very cold, the energy to drive the chemical reactions comes from ultraviolet radiation from the stars — as predicted by theory and confirmed by experiments carried out in the 1980s in which tiny silicate grains covered in just this kind of icy material and kept at a chilly 10 K were dosed with ultraviolet light.

But cold clouds do not tell the whole story of what goes on between the stars. In other places hot clouds of gas shine because the radiation from nearby stars has warmed them to around ten thousand degrees (for such temperatures, it makes little difference whether you are measuring in Kelvin or degrees Celsius), and the radiation from these hot clouds makes it relatively easy to investigate their properties, revealing the presence of many polyatomic molecules and telling us that the clouds contain tens of thousands of atoms in every cubic centimeter — remember that hydrogen atoms still form by far the bulk of the composition of any interstellar cloud, no matter how much it is enriched with stardust.

Overall, interstellar matter contains about 10 percent as much mass as the mass of all the bright stars in the Galaxy put together.[1] Since there are several hundred million stars in the Milky Way, each more or less like the Sun, this puts the mass of interstellar material, at a rather conservative minimum, at 10 billion times the mass of the Sun. This is ample for the making new stars for a while yet — and you have to remember that even if 10 billion new stars were made out of the stuff

1. I am talking here only about the kind of matter that stars, planets, and people are made of, composed of the familiar chemical elements. There is also evidence for the existence of another kind of "dark matter" in the Universe, and the disk of the Milky Way may be embedded in a large spherical distribution of this dark stuff. But it does not play a part in the story I have to tell here.

in interstellar space, that wouldn't exhaust the supply, because interstellar stuff is constantly being replenished and enriched by stellar explosions and mass loss from giant stars. Overall, there must be a more or less steady decline in the amount of interstellar material around, because some of it does end up in the form of white dwarf stars, or neutron stars (or even black holes), and cannot be recycled. But for billions of years past, the process has indeed been one of recycling rather than just using up the original supply of raw material — and will remain so for billions of years to come.

A nice analogy can be made with a large pot of vegetable soup simmering on a stove. It starts out as just water, with one ingredient (maybe carrots) in it. Someone takes a bowl of soup, and in exchange they toss in another ingredient — perhaps tomatoes — and add a little water (but not *quite* as much as the amount of soup they took). As time passes, more people help themselves to the soup and toss something into the pot, but a little bit more stuff always leaves the pot than goes back in. The level of soup in the pot slowly goes down, and eventually the pot will be empty. But along the way the soup becomes richer and richer with complex ingredients, so that it is never the same from one bowl to the next. In a similar way, the very first stars were made only of hydrogen and helium; then some of them exploded and enriched the interstellar medium (in fact, we have never identified any of these primordial stars, which must have all died before even the oldest stars that we see were born). The next generation of stars was made out of slightly enriched material, and the process repeated time and again until today we have stars like the Sun, formed from interstellar material about 4.5 billion years ago, when it had already been enriched by many generations of exploding stars, over a comparable span of time (the Milky Way is a little more than 10 billion years old).

The process is remarkably slow, by human standards. It is estimated that the amount of interstellar material being reworked into new stars each year in our Milky Way Galaxy at the present time is less than about ten times the mass of our Sun. Since most stars are smaller than the Sun, in round numbers you can say that between ten and twenty new stars light up in our Galaxy each year. But in 10 *billion* years, that means that 100 billion solar masses of material, perhaps a third of the mass of all the stars in our Galaxy today, and ten times the mass of the present interstellar medium, has been recycled in this way. All this requires is the ejection of about ten solar masses of recycled material from stars over the Galaxy as a whole each year — either in the form of stellar winds from red giants or in rare supernova explosions — to replace the material turning into new stars. There must also have been a much more intensive burst of activity, in which tens of millions, possibly hundreds of millions, of stars formed at the same time when the Universe was young. We can see such spectacular activity going on in systems known as "starburst" galaxies, sometimes as a result of tidal interactions that occur when two galaxies pass close by each other. But that is getting too far away from my story. Getting back closer to home, one implication of this continuing process of star formation and recycling of interstellar material is that the interstellar medium today is already richer in heavy elements than it was when the Sun formed, so that stars forming today will have a different concentration of chemical ingredients than the Sun. But they are still the same ingredients — stars that formed when the Galaxy was young started out with fewer atoms of heavy elements overall and with, for example, a higher proportion of oxygen, compared with iron, than stars forming today. But they each started out with traces of the same elements.

At last, we are in a position to see how stars like the Sun — and, indeed, the Sun itself — formed. Astronomers are confident that they do understand the basics of this process, not least because we can see it happening today, in a region known as the Orion Nebula, a cloud of hot gas and young stars only about 1,300 light years away from us. As the name suggests, the Orion Nebula is in the constellation of Orion, and it can be seen with the naked eye (just), or more easily with binoculars, as a fuzzy blob in the middle of Orion's sword. Scores of polyatomic molecules have been detected in the Nebula. Because the cloud is lit up by the young stars embedded in it, it makes a spectacular sight on astronomical photographs; but it is only the most visible portion of a so-called giant molecular cloud, which covers almost the entire region of the sky outlined by the constellation of Orion, and has been probed using radio astronomy and infrared astronomy to reveal many hot spots associated with early stages in the formation of stars. Some of the youngest stars in the Orion Nebula itself are only about a million years old, and the conditions there are thought to resemble closely the conditions which existed in the cloud of gas and dust from which our Solar System formed some 4.5 billion years ago (apart, of course, from the fact that the Orion giant molecular cloud is even more enriched in heavy elements, thanks to that extra 4.5 billion years of galactic evolution, than the cloud in which the Solar System formed).

Putting all the observational evidence and all the theoretical models together, it is clear that the Sun formed as part of a giant molecular cloud, which may have contained as much as a million solar masses of material and had a diameter of a couple of hundred light years, which began collapsing about 5 billion years ago. Clouds like this exist all around the disk of a galaxy like the Milky Way, but they are most likely to begin to collapse along the edges of the distinctive spiral arms which are a feature of such galaxies. These spiral features are visible because they are lined with hot young stars. But those stars are there only because the underlying spiral feature is a wave of increased density, sweeping around the disk of the Galaxy. From the perspective of the stars in the disk, each moving around the center of the Galaxy in a roughly circular orbit, you can think of the density wave as a region of high density through which the star passes, rather like a car on a motorway which comes up to a dense region of slow-moving traffic and gradually works its way through and out the other side, leaving the traffic jam behind. Or you can think of it as a flame — in the flame of a cigarette lighter, gas from the reservoir inside the lighter comes up through the nozzle and burns, and the products are dispersed on the other side. The flame itself looks the same all the time it is burning, but actually all the atoms and molecules in the flame are constantly being replaced as they stream through it.

It is the increased density in the spiral arms which gives giant molecular clouds a squeeze, triggering them to collapse and give birth to the hot young stars which then outline the spiral arms, just "downstream" from the region of greatest density. The brightest, most massive stars live fast and die young, never moving far from their birthplaces but spreading their seed back into the interstellar medium; smaller stars, like our Sun, live for billions of years, making many circuits of the Galaxy and becoming widely separated from companions that were born in the same collapsing cloud.

But even with a squeeze provided by the density wave of a spiral arm, the clouds would not be able to collapse all the way down to form stars like the Sun without a little additional help. That help comes from the processed material already in the cloud — in particular, the molecules of water vapor and carbon monoxide present in the gas, together with solid grains of carbon itself.

Our basic understanding of the way clouds of gas collapse and fragment in interstellar space goes back to the work of the British astronomer James Jeans, in the 1920s. If you try to squeeze a cloud of gas, it gets hot, and the heat makes it expand, so that it stops collapsing. Jeans worked out that interstellar clouds will collapse only if they have more than a certain amount of mass, so that once the collapse starts, the gravity of the cloud overwhelms the resistance and it collapses in a rush, breaking up into smaller fragments as it does so. The critical mass necessary for this to happen — known as the Jeans mass — depends on the density of the cloud (in terms of particles per cubic centimeter) and its temperature, which makes the calculations more complicated. And the relationship Jeans worked out is only an approximate description of what is going on, anyway. But broadly speaking a cloud of gas containing, for example, about three thousand times as much mass as our Sun, with a diameter of about 40 light years and a temperature of about 100 K, could collapse down to a diameter of about 10 light years. Because this would increase its density, provided that it still has roughly the same temperature it could break into about 10 pieces, each with a mass about three hundred times that of the Sun, which could collapse in their turn. As the density increases further, each cloud can fragment repeatedly, ending up with objects the size of the Sun and other stars. Stars are born in the densest regions of the cloud, where knots of material form that are too dense for all the radiation to escape. They heat up from inside, which first stops any further collapse and then begins to make them glow as stars.

But one of the key requirements of this whole process is that the temperature stay more or less the same as each cloud is collapsing. The collapse itself generates heat, as gravitational energy is released. So the collapse can continue only if there is a way to lose this heat from the cloud. A cloud can cool down and collapse to the point where stars like the Sun are born only if it has a way to lose energy. Just how it could do this was a great puzzle to Jeans himself, and remained so for the best part of another fifty years, until the rich complexity of the chemistry of such interstellar clouds began to be understood. It is now clear that in the cloud at large, in the early stages of this process of collapse, the cooling is carried out by the molecules of carbon monoxide and water vapor. As they get warm, they radiate in the infrared part of the spectrum. Infrared radiation is very good at penetrating dusty material and escapes out of the cloud entirely, keeping it cool. Then, at a later stage of the collapse, when the first hot stars begin to form, carbon grains come into their own. The first stars, forming in the densest part of the cloud, are massive and bright, radiating a lot of ultraviolet radiation, which would tend to blow the cloud apart and stop any more stars forming — except that it is absorbed by carbon dust in the cloud, and reradiated in the infrared where it can escape more easily into space. With just 1 percent of the mass of the cloud, carbon grains still play a crucial role in allowing many stars to form at once, instead of just a few.

At least, they do today. Clearly, when the first clouds of primordial hydrogen and helium started to collapse, when the Universe was young and the Galaxy itself was just forming, neither of these cooling processes would have been at work. Since no primordial stars from the resulting collapsing gas clouds have survived to the present day, we can only guess at what happened — but the guesswork is backed up by computer models, and the bottom line is that we know those first stars did form, or we wouldn't be here today to puzzle over how they formed. Almost certainly, those primordial collapsing clouds would have had great trouble fragmenting and would have got hot inside, creating massive superstars which ran through their life cycles quickly and exploded, seeding the interstellar material with the first traces of heavy elements. As the heavy elements began to build up (crucially, in this case, the carbon and oxygen atoms), successive waves of star formation would have become easier and easier, as clouds found it easier and easier to radiate away excess heat.

By the time the giant molecular cloud which gave birth to our Solar System began to collapse after being squeezed by an encounter with a dense spiral arm 5 billion years ago, the mixture of stuff in the interstellar medium was 70 percent hydrogen, 27 percent helium, 1 percent oxygen, 0.3 percent carbon, and 0.1 percent nitrogen, with just traces of everything else. Some of the gas in the cloud was in the form of carbon monoxide and water vapor; between 1 and 2 percent of the cloud's mass was in the form of solid grains, about a quarter in the form of carbon, polycyclic hydrocarbons, and iron; and the rest was mainly in the form of iron and magnesium silicates coated in various ices laced with polyatomic organic molecules. It is still only a small proportion of the total mass of the cloud — but remember that the total mass of the cloud is at least a million times the mass of the Sun. Even 1 percent of a million is ten thousand, and ten thousand solar masses of solid grains is equal to more than 3 *billion* times the mass of the Earth. Plenty of raw material there for building new planets.

The first new stars, with masses tens of times the mass of the Sun, would have formed within a few hundred thousand years of the beginning of the collapse of the cloud. These stars are the classic progenitors of Type II supernovae and will run through their life cycles and explode within a few million years, creating a froth of expanding supernova remnants, bubbles of hot gas colliding and interacting with one another. It is these colliding bubbles of supernova debris that encourage the less dense regions of gas and dust in the nebula to collapse and that lead directly to the formation of stars like the Sun and planets like the Earth.

We even have direct evidence that our Solar System formed in this way, as the result of the influence of one or more nearby supernovae on one particular knot of gas some 5 billion years ago. The meteorites that fall to Earth today are leftover fragments of material from the formation of the Solar System, as I shall shortly describe in a little more detail. They represent bits of solid material that condensed, out of the nebula in which the Solar System formed, at the same time as the Sun and planets themselves were forming.[2] The oldest of these meteorites, members of a family known as carbonaceous chondrites, contain little lumps of material rich in calcium, aluminum, titanium, silicon, and oxygen. These blobs often contain unusual abundances of the isotopes of oxygen and magnesium, compared with the proportions of the isotopes of these elements found on Earth, and this provides a clue to the origin of the material. The presence of magnesium-26 is particularly significant, because it is made from the radioactive decay of aluminum-26, which is itself formed in supernovae but has a half-life of only 740,000 years. This means that the magnesium-26 in the chondrites must have been deposited there in the form of aluminium-26 within a few hundred thousand years of the explosion of a supernova, and decayed in situ to make the magnesium-26. It also means that the blobs rich in calcium and aluminum are almost intact lumps of supernova debris, preserved inside the meteorites and unchanged, apart from the changes wrought by radioactive decay, for nearly 5 billion years. Other isotope studies of meteorite samples (and the diamond dust I mentioned earlier) point to the same conclusion — that the material from which the Solar System formed had been hit by the impact of a shock wave of material expelled by a supernova less than a million years before the formation of the Solar System itself.

2. There is a wealth of evidence which points to a time of about 4.5 billion years before the present for the epoch of the formation of these meteorites and the Solar System itself. But I won't go into the details of that here because I did so in my book *The Birth of Time*.

At the same time that this shock wave was triggering the collapse of the particular blob of gas and dust that formed the Solar System (a blob which started out with perhaps twice the mass of the Sun today), it was triggering the collapse of other blobs of gas and dust nearby, and a little farther away still more blobs were collapsing under the influence of other supernova explosions. In the Orion Nebula today, an imaginary cube with sides three light years long would contain thousands of stars, each one, on average, less than a third of a light year from its nearest neighbor. This is just like the conditions under which we think the Solar System formed — yet today the nearest stellar neighbor to the Sun is more than four light years away, so a cube with sides three light years long, centered on the Sun, would not contain any other stars at all. But although a profusion of stars and (possibly) planetary systems emerged from the big squeeze 4–5 billion years ago, from now on I shall focus on the one system that we are particularly interested in, our own Solar System.

Any blob of gas in space is bound to have a tiny amount of rotation, just by chance. As it collapses down into a smaller volume, this makes it spin faster, and the spin (or angular momentum) ensures that material settles into a fat disk around the forming star, instead of all falling into the center. In the early stages of the collapse, the whole ball of gas shrinks by about a factor of a hundred thousand, and given the way angular momentum works, the speed with which stuff is moving around the center of the forming star also increases by a factor of a hundred thousand. By the time a central hot object stabilizes as a glowing red star, containing about 10 percent as much mass as the Sun has today, it is surrounded by the fat disk of dust and gas, while more dust and gas continues to rain down on the disk and the central star from all directions.

Remember that only about 1–2 percent of the stuff in the disk is dust, and the rest gas. But a lot of the gas is going to get blown away by the heat of the young star, forming a kind of stellar wind, while the dust stays in the disk. The disk is fat at first because it is hot — the material falling into it from above and below carries kinetic energy, and this is converted into heat in the swirling disk. The hot gas keeps the dust particles suspended like smoke in the air, battered on all sides by fast-moving molecules. Over an interval of about a hundred thousand years, matter in the inner part of the disk is slowed down by friction and falls into the young star, while matter in the outer part of the disk is accelerated (so that, overall, angular momentum is conserved) and flung away into space. By the time the proto-Sun has reached its present mass, there is no more material falling into the disk, which cools down and settles into a thin layer around the central star, probably in a series of rings reminiscent of the rings of Saturn.

This picture is not based solely on theory and computer models. Many young stars with dusty disks of stuff surrounding them have now been detected and photographed by astronomers, confirming the broad outlines of the story. This is a relatively recent development, since the first such disk, around the star known as Beta Pictoris, was photographed for the first time in 1984; but more than a hundred of these disks had been detected by the mid-1990s. In the archetype Beta Pictoris itself, the disk extends to a distance roughly a thousand times farther from its central star than the Earth is from the Sun, and contains slightly more mass than the Sun. The details of exactly how the dusty material in the thinned-out stages of such a disk surrounding the young Sun formed planets are not yet fully understood, but little more than common sense is needed to get the overall picture in focus. The key factor determining what happened to the dust and icy grains in the disk was the temperature at different distances from the Sun in the early stages of planet formation, and this is fairly simple to work out. At about the distance of the Earth (the planet we are most interested in) from the Sun (a distance known as 1 astronomical unit, or 1 AU), the temperature was around 1000 K, probably a little higher; at 2.5 AU from the young Sun, the temperature was only 450 K, while at a distance of 5 AU it had fallen to around 225 K. The grains in the region where the Earth formed were clearly heated by their contact with the gas in the swirling disk to a point where not only all the icy material covering the grains evaporated, but also the interesting polyatomic molecules, including the organic molecules (the ones containing carbon), were destroyed. But between about 2.5 AU and 5 AU from the young Sun, the temperature was low enough for the organic molecules to remain intact, even though the ices were vaporized. And even farther out, beyond 5 AU, the temperature was so low that ice, including water ice, was preserved as a constituent of the dusty material of the disk. So as the grains of matter in the disk began to stick together and form larger lumps, the end products of this accretion process were different at different distances from the Sun.

The first stage of this process literally involves the stickiness of the tiny grains of dust. Because they are all moving in the same direction around the Sun, when they collide they do so gently, with one grain overtaking another and giving it a gentle nudge from behind. This makes it possible for them to stick together, building up fluffy balls of stuff which before too long contain enough mass to begin to attract one another by gravity. Over a span of about a hundred thousand years, this process builds up objects a kilometer or more in size. These planetesimals would then continue to join together, forming larger and larger objects, while the remaining gas in the disk was swept away by the heat of the young Sun. Within about 2.5 AU from the Sun (the zone roughly out to the present-day orbit of Mars), about a million years after the original blob of gas and dust had started to collapse, there would have been about twenty to thirty objects, each roughly in the range of sizes from the Moon to Mars, together with countless numbers of smaller planetesimals. The bigger objects swept up the smaller ones, and some of them collided with one another, resulting in the four rocky planets we see today in this part of the Solar System — Mercury, Venus, Earth, and Mars. But all these objects were made out of silicate grains that not only had been deiced (and certainly dehydrated) but had been burned free of organic molecules.

Beyond about 5 AU from the Sun, things were different. The same process of accumulation of dust to make larger objects occurred, but there was always plenty of ice. Not only that, a lot of the stuff that had been in the form of ice around grains in the inner Solar System but had evaporated was blown outward by the wind from the Sun, only to freeze again in the outer regions of the Solar System, forming a kind of interplanetary snow which added to the mass of the planets forming there. It is no coincidence that the giant planets Jupiter, Saturn, Uranus, and Neptune, largely composed of gases like methane and ammonia, lie between 5 and 30 AU from the Sun.

Between Mars (which actually orbits the Sun at a distance of 1.5 AU) and Jupiter (at a distance of 5.2 AU), there is a belt of smaller rocky objects, the asteroid belt. They formed in the region where it was still too hot for the ices to survive, but cool enough for interesting organic molecules to be preserved — debris from the asteroid belt still occasionally falls to Earth in the form of meteorites, so their composition has been studied and analyzed in some detail.[3] Far more important for our story, though, in the vicinity of the orbit of Jupiter, and

---

3. The debris in the asteroid belt never stuck together to form a complete planet because of the disturbing influence of the gravity of Jupiter. For more details about the formation of the Solar System and the role of asteroids and comets in all this, see John Gribbin and Mary Gribbin, *Fire on Earth* (London: Simon & Schuster, 1996).

farther out in the Solar System, similar objects formed not just out of rocky material, but out of the variety of ices and snows present, making comets. As Jupiter grew to its present huge mass (318 times the mass of the Earth), its gravitational influence wreaked havoc among these comets, sending many of them hurtling outward into the far reaches of the Solar System (in some cases, out of the Solar System altogether, into deep space) and many of them plunging inward past the inner planets. "Many" scarcely does justice to the numbers involved. The number of comets still present in orbit, far beyond the planet Neptune, in a cloud around the Sun (known as the Öpik-Oort cloud), is estimated at several thousand billion — all ejected in this way in the early stages of the formation of the Solar System. Each comet contains so little mass that together they add up to no more than about three to five times as much mass as the Earth — but the computer simulations suggest that in the early stages of the formation of the Solar System so many comets were deflected by Jupiter into orbits taking them past the inner planets and close by the Sun that their total mass amounted to at least three times as much material, adding up to ten to fifteen times the mass of the Earth. That's a lot of mass, spread out over a vast number of comets. But not all of them actually went *past* the inner planets, or we wouldn't be here now.

It is time to sharpen our focus once again, concentrating now on what happened to the Earth itself as the Solar System formed. As the grains (silicates, iron, and carbon) that were to become a planet started to stick together, at first the process did not involve a great release of energy, because the grains, and later larger lumps of stuff, were moving slowly as they collided with the protoplanet, which grew in size without getting hot. It was only after it had become quite large that its gravitational pull was so strong that it pulled other objects down onto its surface with enough impact velocity (and therefore enough kinetic energy to be turned into heat) that it made the surface hot, and for a long time the core stayed solid, even though the surface of the planet began to melt. The whole process of accumulation of the Earth took several tens of millions of years—perhaps as long as 50 million years—and it was only in the late stages of this process that heat from the surface penetrated to the core and melted it through. With the whole planet in this liquid state (heated by the kinetic energy of impacting meteorites), the dense iron settled down into the core of the planet, while the lighter silicates floated to the surface. As the primordial meteoritic bombardment eased off and came (almost) to a halt, the surface of the Earth cooled and solidified as a layer of mainly silicate rocks, an insulating shell surrounding the still-molten core. Along with iron, the hot, dense core contained smaller, but still significant, quantities of other heavy elements, supernova debris including radioactive uranium. The formation of the insulating layer of solid rock at the surface, trapping heat like a blanket around the Earth, and the energy released by the radioactive decay of uranium has kept the interior of the Earth in a hot, molten state to the present day. Since the energy still being released by that uranium was put into the atomic nuclei when they were made in a supernova explosion (essentially out of gravitational energy released in the collapse of the star that exploded), this means that the Earth's interior is being kept warm today largely by stored-up supernova energy—in much the same way that when burning a lump of coal you are releasing the energy of stored-up sunlight that was locked up in the material that became the coal by photosynthesis in plants that were alive tens of millions of years ago.

Within about 10 million years of the surface of the Earth solidifying, however, the planet experienced the last, cataclysmic step in its formation, when it was struck a glancing blow by an object slightly bigger than the planet Mars (Mars has only one-tenth the mass of the Earth). This remelted the surface layer of the Earth, and any iron brought to Earth by the impact would have melted down into the core of the planet, as the impacting object was completely destroyed in the collision. But the impact also flung a large amount of silicate material back out into space, a mixture of debris from the impacting object and molten silicates from the Earth itself, adding up to perhaps ten times the mass of the Moon (the Moon has only a little more than a hundredth of the mass of the Earth, making it little more than one-tenth as massive as Mars). Most of this material was lost into space; but some formed a ring around the Earth, similar to the ring of material that had formed around the young Sun, but on a smaller scale. The material in this ring coagulated to form the Moon in the same sort of way that the material around the young Sun had coagulated to form the planets. This explains why the Moon has no iron core, and also why the Earth rotates relatively rapidly on its axis, once every twenty-four hours — it is a result of the spin imparted by the collision.

It also explains an otherwise puzzling difference between the Earth and Venus, which is almost a twin to the Earth in terms of size, and presumably formed in much the same way. The solid crust of rock at the surface of the Earth today is relatively thin — only five kilometers thick under the oceans (which account for two-thirds of the surface) and an average of thirty kilometers thick on the continents. Because it is thin, it is relatively easy for the crust to crack, forming plates, rather like the pieces of a jigsaw puzzle, which are jostled by convection currents in the underlying fluid material. This causes continental drift and a continuous stream of volcanic and earthquake activity, particularly in the regions along the cracks between the plates — the whole phenomenon is called plate tectonics.

When space probes sent back data from Venus, though, planetary scientists were surprised to discover no sign of this kind of active surface. There are no plates, and no plate tectonics, on Venus. Instead, by counting the number of craters on the surface of Venus today, and comparing this with the number of craters on the Moon and Mercury, they inferred that the entire surface of Venus had been reworked in a great cataclysm that occurred about 600 million years ago. There are rival models to explain this, but the one I like is the suggestion that Venus has a very thick crust (perhaps fifty to one hundred kilometers deep all over the surface of the planet), so that there is no plate tectonic activity, and heat from the interior is not released steadily by volcanoes. As a result, heat produced by radioactive decay inside Venus builds up over a long period, until the entire surface layer is cracked apart and sinks down into the fluid below, which wells out through the cracks, cools, and forms a new surface.

I like this scenario because it seems to me that there is a natural explanation of why the surfaces of Venus and Earth should be so different, which ties in neatly with the explanation of the origin of the Moon in a collision between the young Earth and a Mars-sized object. Earth has not suffered the fate of Venus because it lost so much of its surface silicate material in that great event and was left with only a thin crust that could easily crack and allow heat from the interior to escape steadily in to space.[4]

This model of how the Moon formed is borne out by many pieces of evidence, including the lack of an iron core and the fact that radioactive dating of Moon rocks tells us that they are slightly younger than the oldest rocks on Earth. Circumstantial evidence also comes

---

4. I should say that this is entirely my own speculation and should not be taken as received wisdom!

from investigations of the planet Mercury, where a major collision seems to have had the opposite effect to the glancing blow that struck the Earth.

Whereas the Moon is like the crust of the Earth without a core, Mercury is like the core of the Earth without any crust. The explanation for this is that early in its life Mercury was also hit by a Mars-sized fragment of planetary debris, but this time it was an almost head-on collision, not a glancing blow. Such an impact would have driven the iron core of the impacting object deep into Mercury, to meld with the core of the original planet. But at the same time it would have blasted away the molten outer layers (melted by the kinetic energy of the impact) of both objects, scattering them completely into space, without giving them a chance to form a ring around the planet and then to coalesce into a moon.

All the evidence tells us that a little less than 4.5 billion years ago the Moon was already in place, orbiting the Earth, and that the Earth was a red-hot ball of rock, gradually cooling down in space. It had no atmosphere, and all traces of water, in particular, had been removed from the material which formed the planet, first as the ice around the dusty grains evaporated in the heat of the energetic disk around the young Sun, and then by the heat of the impacts which formed the planet proper. For the same reasons, there were no interesting organic molecules anywhere on the planet. And yet, the geologic evidence tells us that life had emerged on Earth 3.8 billion years ago. Since that evidence is in the form of fossils preserved in sedimentary rocks — rocks laid down at the bottom of lakes or oceans — it also tells us that there was plenty of water around on the planet by then. What happened, in less than a billion years, to turn an arid, airless desert into a water world that was already a home for life?

During the first 500 or 600 million years of its existence, the

Earth was bombarded by comets ejected from the region of the giant planets. We know, from the laws of mechanics, that much of the icy material from the outer regions of the young Solar System was sent inward by the gravitational influence of the giant planets themselves, and logic tells us that some fraction of the thousands of billions of comets hurtling inward in this way must have hit the Earth and the other inner planets. There is also the evidence of our own eyes — when we look at the battered face of the Moon, we are looking directly at the scars caused by this half-billion-year bombardment of the Earth-Moon system. Radioactive and other dating techniques tell us that this bombardment ended about 4 billion years ago. We also see, with the aid of space probes, similar traces of primordial battering on the surfaces of Mercury and Mars. Venus is a special case, as I have already explained, since there is evidence that the entire surface of the planet was turned over in some great upheaval that occurred about 600 million years ago. The Earth itself is also not as heavily cratered as the Moon, because on Earth geologic processes are continually renewing the crust of the Earth as new material spreads out from cracks in the floors of the oceans, while older oceanic crust is constantly being buried in deep ocean trenches, diving back down to melt into the subsurface layers of the planet. This seafloor spreading is part of the pattern of global tectonic activity that has not only renewed the ocean floor several times in the long history of the planet, but has also set continents moving about the globe, colliding with one another, throwing up great mountain ranges, triggering volcanic activity, and helping to destroy most of the direct evidence of the primordial bombardment.

Although comets have sometimes been described as cosmic snowballs, this doesn't mean they don't release a lot of energy when they hit a planet. The energy released by an impact is the kinetic energy of the object, liberated as heat. This kinetic energy depends only on the mass and speed of the impacting object. From our childhood experiences, anyone who has been in a snowball fight knows that snowballs are not always light and fluffy. A snowball with the same mass as a bowling ball, squeezed hard (like the head of a snowman) and sent rolling down a bowling alley at the same speed as a bowling ball would scatter the pins just as effectively as a regular bowling ball would. The impact energy of a fast-moving comet was memorably illustrated in July 1994, when images of the impacts of fragments of the comet Shoemaker-Levy 9 with the planet Jupiter were a main item of television news for several days. An object the same size as one of those fragments (about ten kilometers across), striking the Earth at a speed of fifty kilometers per second, would release as much energy as the explosion of a hundred million megatonnes of TNT—indeed just such an impact is thought to have been responsible for the "death of the dinosaurs," about 65 million years ago.

When icy material hit the Moon, the energy of the impacts blasted great craters in its surface; but the Moon has such a weak gravitational pull that the gases vaporized during the impact escaped into space. Any that did lie around on the surface would, in any case, be evaporated by the heat of the Sun (except in a couple of very special cases, which I am about to mention). When similar objects hit the Earth, however, although some of the material vaporized in these impacts also escaped, a great deal stayed in the gravitational grip of the planet—some of it condensing out and falling as rain to make the oceans, some staying in the form of gases, building up an atmosphere around the young Earth. Overall, though, the models suggest that ten times more primordial water was brought to Earth in this way than the amount that stayed liquid to form the oceans, and that a thousand times as much gas as there is in the atmosphere today may have arrived in the form of comets. Most of this extra material (generically known as "volatiles") was thoroughly mixed in with the surface layers of the Earth (some astronomers graphically refer to this as "impact plowing"), to form the familiar volatile-rich rocks of the Earth's surface today and to provide a source of the material (such as carbon dioxide) which began to be "outgassed" from volcanoes and recycled through the surface of the Earth (in this case, in the form of carbonate rocks) and back into new volcanoes by the processes of tectonic activity.

In a neat confirmation of this model of the origin of the Earth's volatiles, in the second half of the 1990s space probes revealed the possible presence of traces of cometary ice, including water, at both poles of the Moon — cometary ice which had accumulated in deep, dark craters, in the shadows where the Sun never shines.

The primordial bombardment thinned out around 4 billion years ago (thinned out, rather than stopped, because, as the dinosaurs discovered, impacts from space still happen). Life was present on the surface of the Earth within another 200 million years. The cometary-impact scenario actually reduces the amount of time available for life to emerge and makes the emergence of life an even more dramatic event; but the presence of cometary material in the inner Solar System also explains how life could have gotten a grip on the surface of the Earth so quickly, as I discussed in Chapter 1. Now, we have the background knowledge to appreciate fully the significance of the ideas I sketched for you there.

We already know that the cometary material is rich in polyatomic molecules, including the kind of organic molecules that are the building blocks of life — even things like amino acids, the subunits of proteins. The first claimed identification of an amino acid itself in space was made in 1994, by a team at the University of Illinois. They found spectroscopic evidence (at radio wavelengths) for the presence of glycine, the simplest amino acid, in a cloud of interstellar gas and dust near the center of our Galaxy. The same series of observations also revealed the presence of what the team described as other "large floppy molecules," including ethyl cyanide and methyl formate. Although we have not yet detected other amino acids in space by their spectroscopic signatures, they have been found, along with other complex organic molecules, inside fragments of meteoritic material, rocks from space that have fallen to the Earth. Some of these meteorites are dated to around the time of the formation of the Solar System, at least 4 billion years ago. Although they have only recently fallen to Earth, for the past 4 billion years they have been drifting in space, preserving intact samples of the primordial stuff from which the planets were made. In addition to amino acids (the building blocks of proteins) they contain molecules called purines and pyrimidines, which are subunits of the life molecule itself, DNA. This is direct evidence that such molecules formed in the interstellar cloud from which the Solar System formed and were present in the debris that fell to Earth soon after it formed — if such complex molecules existed in the stony bits of debris floating around the Solar System in those days, they surely also existed in the icy comets that struck the Earth in such profusion then.

It is, though, hard to see how this molecular material would survive the heat of an impact liberating the energy of a hundred million megatonnes of TNT. Partly for this reason, some biologists (and some astronomers) have suggested that life began deep in the bowels of the planet, well below the surface layers which were being plowed up by comet impacts, where the energy of the hot interior could be used as the driving force for chemical reactions in which molecules capable of copying themselves appeared. In some ways, this is a nice idea, because it suggests that life can emerge inside any hot planet. And it does give you an extra 600 million years or so to turn nonlife into life. But there is no need for this nice idea, because by no means all the cometary material arriving at the surface of the Earth is brought there in large objects which generate enormous heat on impact—and by moving the chemical processes that give rise to at least the precursors of life out into space we gain not merely 600 million years, but about 10 *billion* years for those processes to do their work.

This is such an enormous span of time, given the complexity of the molecules that have already been identified in interstellar clouds (and even after allowing for the fact that chemical reactions proceed fairly slowly in those clouds, because of the relatively modest amounts of energy available to drive those reactions), that another, minority view, held by some biologists and astronomers, is that genuinely living systems may have emerged first in space, and only got transferred to the surfaces of planets later. One version of this scenario suggests that there may be cometlike objects in deep space that are warmed inside by the radioactive decay of heavy elements produced in supernova explosions—warmed to the point where they have liquid centers, in which the last stages of chemical development crossing the boundary from nonlife to life may have occurred. Again, this idea has its attractive features, not least the implication that there may be thousands of billions of potential sites for these crucial chemical steps to be taken, even in the comet cloud around our Sun, let alone elsewhere in the Universe. And, like the idea that life originated deep inside hot planets, the interstellar-origin idea implies that life is common in the Universe. But, again like the interior-origin idea, it isn't necessary to go that far to find a plausible way for life to have gotten a grip on the Earth (and presumably on many other planets like the Earth, orbiting other stars like the Sun).

Let's stick to things that we know happened, for certain. As well as the cometary material brought to Earth through major impacts, and heated in the resulting fireballs of those impacts to the point where complex molecules were destroyed, a lot of material fell to the ground more gently. When comets enter the inner part of the Solar System, the heat from the Sun boils away material from their icy surfaces, carrying dust with it, and this material stretches out from the comet to make the tenuous glowing tail (glowing only because it reflects sunlight, not because it is hot) that is the most visible feature of the comet from Earth, even though by far the bulk of the mass of the comet is still in its icy head. Some comets pass many times around the Sun, eventually being disrupted completely by this outgassing and forming a trail of solid material spread out along the original orbit of the comet. If and when the Earth passes through one of these trails of cosmic debris, the larger pieces of the comet dust, the size of grains of sand, produce shooting stars (meteors) when they burn up in the atmosphere. This is why meteor showers recur at certain times of the year, when the Earth crosses the orbits of dead or dying comets — the Leonids, for example, meteors seen every year on 17 November (or a day on either side), are pieces of debris following a comet known as Tempel-Tuttle in its orbit.

Other comets get broken up completely by tidal forces if they pass too close to the Sun, or too close to a large planet — Shoemaker-Levy 9 was disrupted in this way by the gravitational pull of Jupiter, before the fragments of the comet smashed into Jupiter on their next orbit. Either way, as well as the sand-grain-sized pieces of debris that burn up in the atmosphere of the Earth as shooting stars, there is a lot of finer dust, also once frozen inside comets, and before that part of the giant molecular cloud from which the Solar System formed, that settles into the atmosphere more lightly than a feather and drifts gently down to the surface of the planet.

Even today, interplanetary dust particles falling gently to Earth in this way bring about three hundred tonnes of organic matter to the surface of the planet each year. That's *just* organic material — poly-atomic molecules containing carbon — mixed in with even more inor-ganic interplanetary debris. Of course, today there is no prospect of any of this organic material doing anything as interesting as forming new sorts of living molecules; it either gets destroyed by ordinary chemical reactions (particularly oxidation, today — though there was no free oxygen in the atmosphere of the Earth until it was put there by living things) or gets absorbed into the existing chains of life on Earth. But when the Earth was young, and there was neither oxygen in the air to destroy these molecules nor living things around to eat them, the inner Solar System was roamed by many comets, and the amount of organic dust falling to Earth must have been very much higher. A reasonable estimate is that toward the end of the 600-million-year battering of the inner planets by comets, after most of the comets sent into the inner Solar System by the gravitational influence of the giant planets had either hit planets or been broken up in orbit, as much as ten thousand tonnes of organic material fell from the skies (almost literally, manna from heaven) each year. In three hundred thousand years, this would have added up to the mass of all living things on Earth today, and in the couple of hundred million years from the end of the cosmic bombardment to the emergence of life in the fossil record, enough organic material would have fallen to Earth to have made (if it had been spread out evenly) a layer containing twenty grams of organic debris in every square centimeter over the entire surface of the Earth. Twenty grams doesn't sound like much — but remember that a small tub of margarine contains 250 grams of organic stuff. If nothing had happened to the material after it fell to Earth, the depth of the amount of organic debris piled up on Earth in the after-math of the cometary battering would have been equivalent to piling up the entire contents of a 250-gram tub of margarine on every little square 3.5 by 3.5 cm (just under 1.5 by 1.5 in.) over the entire surface of the Earth. Over the entire surface of the Earth, that's an awful lot of margarine, and the cometary debris contained molecules much more interesting than the average tub of margarine does. So, instead of just piling up on Earth, interesting things happened to those molecules, under the influence of the energy available from sunlight and the numerous lightning bolts that crackled through the atmosphere of the young planet.

Even if we don't go as far as accepting the modern version of the panspermia idea, spelled out in Chapter 1, which says that the cometary debris may actually have contained living bacteria or fragments of DNA, the *conservative* interpretation of the evidence is that this debris must have contained all varieties of polyatomic molecules we have identified in space, including things like formaldehyde and the polycyclic aromatic hydrocarbons, and even amino acids. Since we cannot possibly have detected everything there is in interstellar space, it must have contained other chemical ingredients as well. And as a bonus, in the late 1990s a team of NASA researchers carrying out experiments on the kind of mix of materials brought to Earth by comets found that in some cases the conditions in the hot shock wave produced by such an object hitting the atmosphere would encourage chemical reactions in which a mixture of materials including hydrogen cyanide and acetylene (both commonly found in giant molecular clouds) produce the chemical units known as amine groups. As their name suggests, amine groups are components of amino acids; and as I have mentioned, amino acids are the building blocks of the life molecules known as proteins.

I don't want to pretend that anyone yet understands how the building blocks of life first assembled themselves into living molecules. As far as the evolution of life on Earth is concerned, that is still the big unknown. Once the first living cells — essentially the same as some modern bacteria — had appeared, 3.8 billion years ago, the rest was relatively straightforward and is quite well understood. But I do want to end by emphasizing just how well understood everything on the other side of that great unknown is, from the Big Bang itself through the evolution of stars to the interstellar medium in which the clouds that become new stars and planets are produced. In an earlier book, *In the Beginning,* I expressed my surprise that these ideas about the nature of the interstellar medium and the way precursors to life seeded the young Earth almost as soon as it cooled had not received acclaim even in the pages of many science books, let alone popularizations. Seven years later, with even more evidence to hand, it is even more astonishing that the praises of this breakthrough in our understanding of our place in the Universe remain unsung.

To try to hammer the point home, I'll give one last specific example. Formic acid (the stuff some ants squirt out as a defensive weapon, and the stinging ingredient in stinging nettles) and methanimine are two of the polyatomic organic molecules that have been identified in dense interstellar clouds. Together, they combine to form an amino acid, glycine. Glycine itself has now been identified in space, but even if it didn't reach the surface of the young Earth, if both formic acid and methanimine reached the surface of the young planet in the kind of quantities implied by the calculations of how much cometary debris there was around in the newly formed Solar System, it is inconceivable that some of these molecules would not have gotten together to make glycine. And amino acids, it cannot be stressed enough, are just one step away from living molecules.

In my view, one of the most profound discoveries made by science in the twentieth century (indeed, one of the most profound discoveries ever made, and not just by science) is that the Milky Way Galaxy, which is, as far as we can tell, a typical representative of the myriad of galaxies that fill the Universe, is itself packed with the raw materials for life, and that these raw materials are the inevitable product of the processes of star birth and star death. We have answered the biggest question of them all — where do we come from? But hardly anybody outside a small circle of scientific specialists seems to have noticed! Jim Lovelock, the founder of the Gaia hypothesis, is one of the few people to have appreciated fully just what all this means. "It seems," he says, "almost as if our Galaxy were a giant warehouse containing the spare parts needed for life." To be sure, the step from nonlife to life remains at best poorly understood. But it is no mystery where the ingredients of life came from.

I started this book with what may have seemed a metaphor, the idea of life on Earth as stardust, made from material forged inside the stars themselves. I end it with the discovery that this is not a metaphor at all — it is the literal truth. The raw material from which the first living molecules were assembled on Earth was brought down to the surface of the Earth in tiny grains of interplanetary material, preserved in the frozen hearts of comets from the interstellar debris of the giant molecular cloud from which the Solar System formed. Those grains themselves literally, not metaphorically, formed from material ejected by stars. The "manna from heaven" that carried the precursors of life down to the surface of the Earth was literally, not metaphorically, stardust. And so are we.

Appendix

Across the Universe(s)

The story I have told in this book explains the relationship between life and the Universe, from the Big Bang to the arrival of molecules of life on the surface of the Earth. It is a complete and self-consistent story, describing our cosmic origins from stardust. But it is not necessarily the whole story of life and the Universe, and in this appendix I want to describe briefly one of the more intriguing current ideas that may, if proved correct, take us beyond the story so far — but with the proviso that "intriguing" doesn't necessarily mean "correct."

It is the suggestion that there may be at least an analogy (and perhaps much more than an analogy) between the origin and evolution of the entire Universe and the origin and evolution of a living organism. These ideas formed a major theme of my earlier book *In the Beginning*, but that was published in 1993, and there are aspects of the story that deserve updating as we enter a new century.

The starting point for these ideas is the discovery that many of the properties of the laws of physics seem to be remarkably fine-tuned to make the Universe a suitable home for life as we know it. Take, for example, the relationship between the four forces of nature that affect the behavior of fundamental particles. Three of these, electromagnetism and the two nuclear interactions, differ greatly in their strengths, but they are all enormously stronger than gravity, the weakest of the four. To put gravity in perspective, the electrical force of repulsion between two protons is about 10^{38} times bigger than the gravitational force of attraction between two protons, so it is no wonder that the electrical force completely overwhelms the gravitational force. It is because gravity is so weak that stars are so big — it takes the gravitational contribution of a very large number of particles (about 10^{57} in the case of the Sun, all protons and neutrons) to crush the material in the heart of a star to the point where electrical repulsive forces are overcome and nuclear fusion can occur. But if gravity were just ten times stronger than it actually is (still only 10^{-37} times as strong as electromagnetism), things would change so that nuclear fusion would be a lot easier, and the lifetime of a star equivalent to the Sun would be only 10 million years, not 10 billion years. That would not be long enough to allow evolution to proceed on planets in the way that it has on Earth.

It is reasonable to speculate on what the Universe might be like if such changes were made to the laws of physics, because we have no idea why things like the forces of nature have the values they do. And there are many "coincidences" of this kind. The one which I find most intriguing I have already commented on — the existence of the carbon resonance that allows the triple-alpha process to take place inside stars, and the corresponding coincidence that the equivalent oxygen resonance is at just the wrong level to allow all the carbon to be promptly converted into oxygen. The appropriate energy level in carbon is just at the limit at which the triple-alpha process can occur; the equivalent oxygen resonance is simply too high for carbon and helium to combine all in a rush.

As I say, there are many coincidences of this kind, and I won't go into them all here;[1] some people suggest about twenty examples of the fine-tuning of the laws of physics which make life forms like us possible. All twenty are needed if we are to exist. And in all those cases, we have no a priori reason why the laws of physics should have these particular values.

One way of looking at this is to say that these are not coincidences at all, but a kind of tautology. We have evolved in a Universe with a certain set of physical laws, so it should be no more surprising to find that our evolution has taken advantage of those conditions than it is to discover that polar bears have evolved thick fur to keep them warm, while monkeys are adapted to a life in the trees. We are what we are, the argument runs, because the Universe is what it is. But there is a school of thought which holds that the Universe did not have to be the way it is — that other laws of physics could have emerged out of the Big Bang. This is like asking how life on Earth might have evolved if there were no ice at the poles and no tall trees at lower latitudes. Would we still have polar bears and monkeys? And this is where the idea of cosmic evolution comes in.

The key word here is "evolve." The new, and admittedly still speculative, way of explaining the cosmic coincidences is to draw an analogy with the evolution of life on Earth. It comes from Lee Smolin, a physicist based in New York, Andrei Linde, a cosmologist based in California, and a handful of other researchers. Their thesis is that the way the Universe works can best be understood not simply by

1. But see John Gribbin and Martin Rees, *The Stuff of the Universe* (London: Penguin, 1995).

applying the rules of physics worked out by Isaac Newton and Albert Einstein, but by taking account as well of the rules of evolution worked out by Charles Darwin and Alfred Wallace — the theory of natural selection. The Universe itself may literally be alive, in this picture, and, more to the point, may have evolved by natural selection from a simpler state to produce the complexity we see around us.

Taking the equations of the general theory of relativity at face value (and nobody has ever found an experiment or observation which suggests that we shouldn't do so), the Big Bang itself emerged from a point of infinite density, a singularity. There is another place where singularities are calculated to occur, using that same theory of relativity — at the heart of a black hole. Roger Penrose and Stephen Hawking proved, in the 1960s, that the expanding Universe is described by exactly the same equations as a collapsing black hole, but with the opposite direction of time. If all the complexity of galaxies, stars, planets, and organic life has emerged from the singularity in which our Universe was born, within a black hole, couldn't something similar be happening to the singularities at the hearts of other black holes?

A naive guess of what might happen to something collapsing *into* a singularity to turn it into the kind of expansion *from* a singularity that we see in our Universe is that there is a "bounce" at the singularity, turning collapse into expansion. Unfortunately, that won't work. A singularity forming from a collapse within our three dimensions of space and one of time cannot turn itself around and explode back outward in the same three dimensions of space and one of time. But in the 1980s relativists realized that there is nothing to stop the material that falls into a singularity in our three dimensions of space and one of time from being shunted through a kind of spacetime warp and emerging as an expanding singularity in another set of dimensions — another spacetime.

A baby universe can be thought of as a pinching off from the bubble of space-time represented by our Universe, with the two connected by a wormhole.

Mathematically, this "new" spacetime is represented by a set of four dimensions (three of space and one of time), just like our own but with all of the new dimensions at right angles to all the familiar dimensions of our own spacetime. Every singularity, in this picture, has its own set of dimensions, forming a bubble universe within the framework of some larger spacetime.

One way to picture what this involves is to use an old analogy between the three dimensions of expanding space around us and the two-dimensional expanding surface of a balloon that is being steadily filled with air. The analogy is not with the volume of air inside the balloon, but with the expanding skin of the balloon, stretching uniformly in two dimensions, but curved around upon itself in a closed surface. Imagine a black hole as forming from a tiny pimple on the surface of the balloon, a small piece of the stretching rubber that gets pinched off, and starts to expand in its own right. There is a new bubble, attached to the original balloon by a tiny, narrow throat — the wormhole. And this new bubble can expand away happily in its own right, to become as big as the original balloon, or even bigger, without the skin of the original balloon (the original universe) being affected at all. There can be many bubbles growing out of the skin

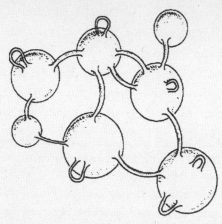

If one baby universe can form from a black hole, then there could be an enormous number of universes (in principle, infinitely many) connected by a complex web of wormholes. This is the basis for speculations that universes themselves may evolve, in the Darwinian sense.

(the spacetime) of the original universe in this way at the same time. And, of course, new bubbles can grow out of the skin of each new universe, ad infinitum.

The dramatic implication is that many — perhaps all — of the black holes that form in our Universe may be the seeds of new universes. And, of course, our own Universe may have been born in this way out of a black hole in another universe. This shifts our view of the Universe by suggesting that the Universe is not unique. Instead, it is one of a population of universes, interconnected by what physicists call wormholes, all of which are rivals for room to grow in multidimensional spacetime.

Smolin's idea is that every time a black hole collapses into a singularity and a new baby universe is formed, the laws of physics themselves are altered slightly as the new universe emerges from the wormhole, in just the way genetic variability among organic life forms on Earth makes offspring slightly different from their parents and provides the raw material for evolution by natural selection.

If the random changes in the workings of the laws of physics — the mutations — happen to allow a little bit more expansion, a baby universe will grow a little larger. The larger it is, the greater the chance for it to make new black holes, to make new singularities, and thereby to trigger the birth of new universes.

Those new universes will also be slightly different from their parents. Some may lose the ability to grow larger and will fade away without leaving offspring. But some may grow even larger than their parents, producing more black holes and giving birth to more baby universes in their turn. The number of new universes that are produced in each generation will be roughly proportional to the volume of the parent universe. There is even an element of competition involved, as the many baby universes are in some sense vying with one another, jostling for elbow room in spacetime.

Heredity is an essential feature of life, and this description of the evolution of universes works just as if we were dealing with living systems (Smolin would say that this is because we *are* dealing with living systems). In this picture, universes pass on their characteristics to their offspring with only minor changes, just as people pass on their characteristics to their children with only minor changes.

Universes that are "successful" are the ones that leave the most offspring. Provided that the random variations are indeed small, there will be a genuinely evolutionary process favoring larger and larger universes. In succeeding generations of universes there will be a natural evolution toward a drift in the laws of physics to favor the production of the kinds of stars that will eventually form black holes. The end product of this process should be not one but many universes which are all about as big as it is possible to get while still being inside a black hole, and in which the laws of physics encourage the formation of stars and black holes. Our Universe closely matches that description.

Smolin is particularly fond of pointing out that the presence of carbon and oxygen in the Universe, which depends on the fine-tuning of those nuclear resonances, is a key factor in the process of star formation and the formation of black holes, not just in the evolution of life. New stars form from clouds of gas and dust in space only because those clouds are able to cool, radiating away heat as they contract. And one of the main reasons they are able to cool is that they contain carbon monoxide, which radiates away energy in the infrared part of the spectrum. This kind of argument suggests that the people who say that life forms like us exist because the Universe is the way it is are right — life makes use of carbon and oxygen because it is there. But it is there because the Universe has evolved to be good at making stars and black holes!

This kind of argument can be applied to all the other puzzling coincidences in the laws of physics, to explain the mystery of why the Universe we live in should be the way it is. You would not expect a random collection of chemicals to suddenly organize themselves into a human being, and in the past this has led some people to seek a supernatural explanation for our existence. But the idea of evolution by natural selection removes the need to invoke the supernatural. In the same sort of way, you would not expect a random collection of physical laws emerging from a singularity to give rise to a Universe like the one we live in. This realization has led some people to suggest that the Big Bang itself may have resulted from supernatural intervention. But evolution by natural selection can also remove the need to invoke the supernatural where the whole Universe is concerned. According to Smolin and Linde, we live in a Universe which is the most likely kind of universe to exist.

But there is no suggestion that the Universe has evolved in that way specifically because it suits life forms like us; rather, it has evolved to produce black holes, and life has taken advantage of the conditions that favor the production of black holes. In a sense, life forms like us are parasites, feeding off the processes that manufacture black holes. You shouldn't find this too new or shocking an idea. After all, life on Earth depends on a supply of energy from the Sun, energy which comes, ultimately, from nuclear-fusion reactions going on at the heart of the Sun. Those nuclear processes do not go on for our benefit, and in the same sense we are parasites feeding off the flow of energy produced by those reactions.

All this does, though, depend on the claim that our Universe is set up (or has evolved) in such a way that it is just about as efficient at producing new black holes, and therefore new universes, as it is possible to be. The arguments on this point get quite technical, and I won't go into them here. But critics of Smolin's approach claim that it would be possible to fine-tune the laws of physics to make the Universe even more efficient at producing black holes (and therefore baby universes). They argue that in that case evolution ought to have done the job already, if Smolin is right, and since the Universe is not perfect at making black holes, therefore he can't be right. I have my doubts about this argument—after all, evolutionary fine-tuning has undoubtedly been at work here on Earth, but that hasn't resulted in any species yet being perfect, including our own. Smolin counters each of these arguments as they are put forward with his own, and so far he has successfully refuted all of them. As a result, over the past few years his ideas have gained in strength, as each apparent loophole has been plugged. But even if someone finds a loophole that cannot be plugged, there is another possible explanation for why the Universe is what it is.

Edward Harrison, of the University of Massachusetts, has developed an idea also put forward in a less worked-out fashion by other cosmologists (notably Alan Guth, of MIT). He points out that it is entirely possible that our Universe was indeed created—not by God, but by intelligent beings with a level of technology only slightly more advanced than our own.

Remember that in the baby-universe scenario every black hole gives birth to a new universe. To make a universe, all you have to do is make a black hole. We do not have the technology to do this, but we do have the scientific knowledge to understand how it might be done—by squeezing a lump of matter to very high densities. And although we don't know of any way to communicate between universes, that is not to say that beings capable of actually making them might not have worked out a way of looking at what goes on inside their creations. They might make universes with different sets of physical laws just because it is possible to do so, and they want to study them; it might even be possible that the creation of universes provides them with a resource that they need, for some purpose that is incomprehensible to us. Or even for a comprehensible purpose— one obvious possibility is that a supercivilization might be able to extract energy from black holes. Or it could even happen by accident, in the way described by Gregory Benford in his superb novel *COSM*. But that really is wandering into the realms of science fiction, and almost time to call a halt. There is, though, one comforting aspect of Harrison's variation on the theme, for anybody who objects to being called a parasite. In the very near future, on the cosmic timescale, our species may be manufacturing black holes, and therefore baby universes, in the same way as Harrison's hypothetical supercivilizations.

In that case, we will be helping our Universe to reproduce itself, and that would mean raising our status from that of a mere parasite to that of a respectable symbiote—a partner (if not quite an equal partner) in a marriage of convenience.

Nothing in this appendix should be taken too seriously. But the rest of the book is intended entirely seriously. Whatever the reason(s) for the laws of physics being the way they are, there is no doubt that the Universe is set up in such a way that the production of carbon, oxygen, and nitrogen in profusion (by human standards) is an inevitable consequence of the life cycles of stars, and that it is inevitable that planets like the Earth will form around stars like the Sun and be laced with complex organic molecules, originally from interstellar clouds, by the arrival of comets. We are made of stardust because we are a natural consequence of the existence of stars, and from this perspective it is impossible to believe that we are alone in the Universe.

Further Reading

Most of the books listed below provide additional information about topics touched on in this book, at about the same level of accessibility. The ones marked with an asterisk are slightly more technical.

Amir Aczel. *Probability 1*. New York: Harcourt Brace, 1998.

Gregory Benford. *COSM*. London: Orbit, 1998.

Francis Crick. *Life Itself*. London: Macdonald, 1982.

George Gamow. *The Birth and Death of the Sun*. New York: Viking, 1940.

John Gribbin. *Almost Everyone's Guide to Science*. New Haven: Yale University Press, 1999.

——. *In Search of the Double Helix*. London: Penguin, 1995.

——. *In the Beginning*. London: Viking, 1993.

John Gribbin and Mary Gribbin. *Fire on Earth*. London: Simon and Schuster, 1996.

Fred Hoyle. *Home Is Where the Wind Blows*. Mill Valley, Calif.: University Science Books, 1994.

William Kaufmann. *Stars and Nebulas*. New York: Freeman, 1978.

——. *Universe*. New York: Freeman, 1988.

Rudolf Kippenhahn. *100 Billion Suns*. London: Weidenfeld and Nicolson, 1983.

*Stephen Mason. *Chemical Evolution*. Oxford: Oxford University Press, 1991.

*A. J. Meadows. *Stellar Evolution*. Oxford: Pergamon, 1978.

I. S. Shklovskii and Carl Sagan. *Intelligent Life in the Universe*. New York: Holden-Day, 1966.

Lee Smolin. *The Life of the Cosmos*. London: Weidenfeld and Nicolson, 1997.

*John Stares. *The Chemistry of Life*. London: Chapman, 1972.

*Roger Tayler. *The Origin of the Chemical Elements*. London: Wykeham, 1972.

Steven Weinberg. *The First Three Minutes*. London: Deutsch, 1977.

Acknowledgments

I am grateful to Virginia Trimble for reading and commenting on the entire text; although I did not always take her advice, this feedback considerably improved the historical presentation of my story. Thanks also to Jonathan Gribbin for his usual efficient work on the illustrations for this book.

Index